The Voice of the Customer for Product Development

Your illustrated guide to obtaining, prioritizing and using customer requirements and creating winning products

4th Edition

By: José Campos and Jean-Claude Balland

Published by:

Rapidinnovation Press
www.Rapidinnovation.com

The Voice of the Customer for Product Development (4th Edition)

By: José Campos and Jean-Claude Balland

Published by:
Rapidinnovation Press
Rapidinnovation, LLC
25000 SW Big Fir Rd.
West Linn, Oregon 97068
United States
www.rapidinnovation.com

ISBN-13: 978-0-9817595-2-4
ISBN-10: 0981759521

Published in the United States

CIP data available from the publisher

Get your interactive PDF version FREE!
See the last pages of this guidebook for details

V-VOC/082712/6.1.2/ACS/PRN/SW

Introduction

Understand What Customers Want

Faced with a Need

The Master Process

Phase I: Investigate

Step 1: Plan Your Visits

Introduction

The Long Journey

Critical Tools

Deployment of a New Tool

Yellow Sticky Protocol

Affinity Diagrams

Prioritizing Affinity Diagrams

Karnaugh Maps

Process Mapping

Customer Panels

Appendixes

Appendix A: Recommended Books

Appendix B: Glossary

Appendix C: Life Cycles Models

Appendix D: Templates

Introduction

Legend

Navigate through this Guidebook

To ease navigation, we have provided a set of icons, which provide orientation and clarity. They are consistent throughout the guidebook, always carrying the same meaning.

Symbols to Indicate Facts to Know

Important Suggestion

Used with each important suggestion, note, or tip, derived from experience in the field during actual projects.

Essential Quote

A quick way to read very important information. We suggest you read all essential quotes.

Flexibility

Indicates the flexibility of a particular tool. We encourage you to adapt the tool to your specific situation and not be governed by rigid, prescriptive frameworks.

Definition

A defined term. See also our Glossary at the back of the guidebook.

Time Frame

To help you plan the likely duration of a meeting, work session, or other event.

You Are Here

One more way to get oriented or to check that you have take the proper steps up to this point.

Symbols for Links to Helpful Info

Note for Print Version

Interactive features are available in the electronic PDF version. If you have a printed version, contact the authors to learn how to upgrade your guidebook.

Template (Printed Version)

You will find copies of the templates at the back of your guidebook.

Template (Electronic Version)

Double click this icon to open a corresponding template (enabled for use in Adobe Reader and Acrobat) that you can fill with your own information and share with your peers and colleagues. When you close the template, you will return to this guidebook.

Each time you open a template from your guidebook, it will be fresh. But you can save versions that you have completed, partially filled in or left blank for future access. Saved versions remain enabled for revisions and collaboration.

These enabled templates also provide you with Comment & Mark-up Tools. Just as you might with a hard copy, you and your team can highlight text, add notes and make custom annotations.

See Also (Electronic Version)

Clicking this icon will jump you to a place in the guidebook with more information relevant to the topic at hand.

Back Button (Electronic Version)

After following a link, you can easily click to return to your original place in the guidebook.

Foreword

By Edward F. McQuarrie

The central task of every customer interview is to understand what the customer values. You cannot sustain a business unless you offer something for which customers are willing to pay. So the logical thing to do would seem to be, just ask.

Alas, there is more to the concept of a "customer visit" than meets the eye. For example, talking to the wrong person will not help you, no matter how skilled an interviewer you might be. So, the first and crucial step in executing a set of customer visits is simply to ask, Who should I visit?

After selecting the proper customers to interview, the next step is to ask the right questions, in the right fashion, and at the right time. To do this, you must craft a solid discussion guide that is responsive to your business objectives. Next, you have to approach the interviews themselves in the right spirit, with an open, questioning, responsive frame of mind. If you don't listen, you can't learn.

Finally, once you are done with all the interviews you must identify and flesh out key customer requirements. For this, you must have a conceptual framework, lest you drown in the details, and lose sight of what the customer really requires. You are fooling yourself if you believe that writing down what the customer says he wants, and then building exactly that, is going to lead to market success. You have to get at the underlying need, which may never have been vocalized, but informs everything the customer said.

In short, "customer visits" requires a structure to guide you and your team through the entire process: from defining the purpose, to selecting the right customers to interview, to identifying and prioritizing customer requirements. You have to execute against all of these if you want to enable your R&D team to develop innovative solutions that deliver value to customers.

More and more companies have come to believe that the die is cast at the very beginning, during product definition, where customer value is uncovered, identified and prioritized. Executives expect the marketing organization to play the focal role in insuring that the product delivers superior value. Marketing needs to, and ought to, take the lead in identifying opportunities to create value for the customers. Marketing is also responsible for prioritizing requirements relative to competitor offerings, company profit goals, and the intensity of customer need. Differentiation is key: you've got to deliver superior value. In a fast-moving global marketplace, it is imperative to have a roadmap that can guide you through the critical steps in identifying, prioritizing and using the voice of the customer.

Developing a deep understanding of your customer, and translating this understanding into a well-differentiated product, requires a plan. The four-phase framework developed in this guide shows you how. The "Four I's" is an innovative and useful way to understand the four distinct stages that your team must transit to deliver innovative products to your customers.

I believe that Jose Campos and Jean-Claude Balland have made a significant contribution to addressing these challenges with the introduction of Version 4.0 of the Voice of the Customer for Product Development Guide. The focus on detail, the ease of navigation, and the rich content make this new edition a valuable addition to the stock of knowledge on how to discover what customers value. The authors' extensive experience informs every page. It is an exceedingly practical guide.

Edward F. McQuarrie

Professor of Marketing, Santa Clara University and author of Customer Visits and The Market Research Toolbox.

Introduction

Approach

Why We Wrote this Guidebook

"We have only two sources of competitive advantage: 1. The ability to learn more about our customer faster than the competition, and 2. The ability to turn that learning into action faster than the competition." — Jack Welch

This quote from Jack Welch says it all. The ability to learn about customer requirements is the soil in which successful and innovative new products grow. How to gather good requirements is what this guidebook is about, and its originality is twofold:

1. Providing you detailed processes and tools with step-by-step guides, which you will be able to use as a "How To" manual to plan and execute a customer requirement program.
2. Focussing you on the search for creating customer value.

A How-To Manual

Having been on the receiving end of guidebooks, handbooks, "primers" and training during our lives in the corporate world, we realized that often the information was not sufficient for us to actually use it in our situation, or the information was formatted poorly such that it made it impossible for us to apply it, which resulted in much frustration, not to mention loss of productivity. Worse yet, the documentation we received from training sessions was simply a hard copy of the PowerPoint slides shown during the session, which were of little help to us when it came time to apply the learning.

Essential

"We have only two sources of competitive advantage: 1. The ability to learn more about our customer faster than the competition, and 2. The ability to turn that learning into action faster than the competition." — Jack Welch

We made a commitment that our guidebook would contain everything a professional would need to apply every concept in it. We have strived to provide the tools, methods and encouragement for you to successfully understand and apply the voice of your customers.

A Search for Value

The second goal was to provide a framework for identifying opportunities to create customer value. While the word "value" is widely used, few product team people go beyond the generic expression, "we need to create customer value." Many use the expressions "value-based," "value-driven" and "value-centric," but few can articulate specifically how their products or companies create value. It is the ambition of this guidebook to help product team people understand the various ways of creating customer value, and to equip them to actually plan and execute a requirement-gathering campaign centered on defining products that can deliver superior customer value.

How to Use this Guidebook

We developed this guide for individuals and teams who are in product development (Marketing and R&D) and have specific assignments to deliver results. The entire focus of the guidebook is to lead you — step-by-step — through each and every concept, even if you have never performed the task before.

The sequence of this guidebook is based on a chronological approach — the outline and format follows a logical "starting point" and move towards a "end point." This means that the higher the page number, the later in the process.

This is not a book about concepts. We have limited concepts to the minimum necessary to create context or perspective. The focus of this guidebook is on the "how-to."

The Process Maps

The most important element of this guide is the process maps — the step-by-step notes for starting and completing key tasks. These process maps lead to the various tools and templates.

There is a hierarchy of process maps:

LEVEL 1: THE MASTER PROCESS This is the highest view of the process. The Master Processes, also known as Mental Models, represent big pictures, and they are rarely actionable due to their strategic nature. They are simply here to give you the overall perspective of a large endeavor or system.

LEVEL 2: MAIN PROCESS These are the major phases contained in the guidebook. They provide a phased (stage) approach and are indeed actionable. Main Processes as generally good for planning purposes, also to discuss with your team as you prepare to start a particular stage.

LEVEL 3: DETAILED PROCESS These are the processes you will need to start and complete tasks. They not only contain the steps, but we have also included critical information in the form of tips, suggestions to help you locate the progress and other important information, please see the Legend at the start of this guidebook for further details on the icons used to help you navigate.

The Template and Tools

Templates are forms that contain information that you and your team enter in order to collaborate and create archival information for later use. Tools are matrices, spreadsheets, tables and other instruments.

It is important to note that for every tool or template there is at least one example and in some cases several. This is intended to provide you with a general idea of the information that should be entered. It is not intended to limit what you and your team feel needs to be done.

Essential

"The most important element of this guide is the process maps — the step-by-step notes for starting and completing key tasks."

Unique Capability for the Electronic Version

Not only does every template in this book have a full example, but also it is fillable. You and your team can enter information about your project and share it. You will notice a link in the form of an icon and when you click on it, you will be able to open a blank template that you and your team can use. Once you have completed entering the information, the template becomes its own PDF independent of the guidebook. You can save it using your own name or title, which you can also share with the rest of your team around the world. You can continue entering information until you are satisfied.

Flexibility

We understand that every development project is different — by definition. Consequently, our intent is not to limit your flexibility. All the process maps, templates and tools can be modified to better meet your objectives and style. We suggest that the first time you apply the these process maps, templates and tools that you try to adhere to our format. As you become familiar with the intent, you can modify them until you have something that is unique to your organization and also meets your requirements to the fullest. Don't be afraid to experiment and modify.

Mission of this Guidebook

The mission is to create awareness of the framework and motivation needed to successfully understand customer needs, process the information and put it to use to develop successful new products. Additionally, this guidebook will provide the tools, templates and methods for obtaining, processing and using customer input.

This guidebook will introduce you to a customer-driven methodology. It will introduce you to 10 steps to increase the value of your customers' offerings. You will learn how to identify opportunities by interviewing customers. You will also learn how to transform these opportunities into requirements and product concepts that cast an optimum balance between usefulness, usability and desirability.

Introduction

Objectives of this Guidebook

The overall objective is to create the capacity for improved productivity! And the results of using this guidebook will include these specific objectives:

- You will be familiar with overall strategies (approaches) needed to incorporate the voice of the customer into new products and services.
- You will understand the ways and means needed to obtain, process and incorporate the voice of the customer into new products and services.

Your Input

No book is perfect; this one is no exception. If you find difficulties in using the tools, concepts and processes, do not hesitate to contact the authors: voc@rapidinnovation.com..

Related Guidebooks for Product Development

In addition to *Risk Management and FMEA*, our library includes several related guidebooks in the Product Development series:

- Risk Management and FMEA for Product Development
  Your illustrated guide to reducing time-to-market through risk management and FMEA.
- Actionable Metrics for Product Development
  Your illustrated guide to developing and using metrics to improve product margins and reduce time-to-market.
- Debriefs and Postmortems for Product Development
  Your illustrated guide to improving performance through lessons learned.
- Flexible Project Management for Product Development
  Your illustrated guide to making project management work in fast-changing environments.

Training, Consulting and Facilitation

Our highly experienced authors are also available to provide direct consulting to your organization. They can be team facilitators to help you implement the contents of these guidebooks. They can provide training on the topic plus many other areas of product development. Contact us for more information.

For comments, questions or additional copies please e-mail us: jose@rapidinnovation.com.
Visit us at www.rapidinnovation.com.

Get Your Interactive PDF Version FREE!

See the last pages of this guidebook for details on how to acquire your free copy of the interactive PDF version.

About the Authors

Jean-Claude Balland Ph. D, JCB Associates

Jean-Claude is a high-technology marketing executive, consultant, professor, and lecturer with extensive experience in sales and marketing management in Europe, the United States, and Japan for technology products. Jean-Claude helps companies develop breakthrough offerings and strategies. He is an adjunct professor at OHSU School of Engineering and Computer Sciences and at Portland State University where he teaches Technology Marketing and other courses related to creating breakthrough products. Prior to his present occupation, Jean-Claude was Director of Internet Marketing at Mentor Graphics Corporation. He worked nineteen years at Tektronix, Inc. and left after several years as the Division Marketing Manager of a large and highly successful division.

José Campos, Principal Consultant and Founder, Rapidinnovation

José collaborates with B2B technology companies around the world on the methods, tools and cultural shifts to rapidly and consistently deliver superior value to customers.

José has spent his career working to define and deliver innovative products to market. For 30 years he has passionately focused on discovering new methods and approaches to discover and deliver superior solutions to the customer in the tumultuous world of high-tech

He has written several acclaimed books: *Voice of the Customer for Product Development* — a methodology to discover, prioritize, and use customer requirements to create clearly differentiated products — faster than the competition. Also, *Risk Management for Product Development* — the strategies and methods to reduce the impact of uncertainty on your investment. Jose was coauthor of Project Management Toolbox in collaboration with Dr. Dragan Milosevic.

His focus on high-tech and his technical background allow him to provide meaningful and practical coaching to technology professionals. He has traveled the world help R&D and Marketing teams improve their ability to create innovative products that customers love.

Understand What Customers Want

Overview

This section will introduce you to a highly important concept in marketing: **VALUE**. Everything customers do (and for that matter, everybody does) is because, "It's worth it." That's a very simple definition of value. To identify what "worth it" means, we will go deeper and introduce several concepts to allow marketers to probe customers in areas of potential added value.

This is a conceptual section that will be used in all subsequent sections and in all efforts to better understand customers.

You will discover that value can be created by developing benefits in five areas and reducing costs in as many areas. From this, you will see how to assess the value your company and its competitors provide with a tool called a Value Profile™.

Before you turn the pages of this chapter, prepare yourself to think that value can be created by other approaches than just providing better specs and/or lower prices. Many of the most important breakthroughs come from precisely understanding these dimensions of value that your competitors ignored.

Faced with a Need

All You Have is 12 Hours

Think about the last time you defined a new product. How much time did you spend with customers? How many customers did you interact with? Was it enough? Did you get all you needed? Consider this: how well your new product fits your customers' needs depends on the time you spend with your customers. On average, product teams define new products based on 12 one-hour sessions with customers. This guidebook introduces you to a customer-driven methodology and shows you how to get the most return out of these 12 hours — how to get better insights into what your customers value and how to transform this understanding into requirements and concepts that deliver real customer value.

Faced with the continual need to introduce new products, a company can take two approaches. The first and most common, is the "faster, smaller, cheaper" approach, based on continuous improvement of the initial concept. This is a technology, or product-driven, approach. But how fast is fast enough? How small is small enough? How cheap is cheap enough? It is easy for a company to overshoot the technical requirements without even realizing it. Intel — and the entire semiconductor industry — recently faced such a critical milestone and realized that after years of increasing microprocessor speed, customers' basic needs were fulfilled and they needed something else. But what? Suddenly, technologists and engineers could not provide the answer. Actually, they had too many and no ways to decide which direction would be the best.

At this juncture, companies facing such a challenge must turn toward the customer to define direction. That's easier said than done. Rediscovering the customers and learning to listen to their needs is difficult after years of telling them what is good for them. But failure to do so is not an option because the penalty is an accelerated loss of market share.

Essential

"This guidebook introduces you to a customer-driven methodology and shows you how to get the most return out of these 12 hours"

In addition, the accelerated product development cycle now makes it possible for any new features to be rapidly copied. So any technical advantage can rapidly evaporate. There is only one solution: continuous innovation. To be successful, these innovations need to be embraced by the customer. And to be embraced by the customer, these innovations need to deliver value. Therefore, companies need to change their product-driven process to one that gives customers a more central role in driving the definition of innovative products.

Why Visit the Customer

We assume you're already convinced of the importance of customer interaction. We also assume you're reading this guidebook because you're ready to have useful customer interaction. But before you begin, take a moment to consider the dramatic difference between your world and that of your typical customer.

The separation between you and your customer is very wide indeed. Consider that the only thing you may have in common is your product — everything else is remarkably different.

Your World	Your Customer's World
What trade publications do you read?	What trade publications does your customer read?
How much do you care about your product?	How much does your customer care about your product?
What are your goals and objectives?	What are your customer's goals and objectives?
What is your level of education?	What is your customer's level of education?
What is your corporate culture?	What is your customer's corporate culture?

Your customer is focused on solving his problems, living his life. He spends little time thinking about your company and your products. This is why it's critical to understand the world of your customer.

Faced with a Need

Customer Value, Benefits and Costs

> **Definition**
> **Customer Value**
> The value of an offering for a particular person is the perception he or she has of the difference between the benefits he or she will derive from the buying and use of this offering under a particular situation, to achieve specific objectives, and the costs he or she will have to pay to get these benefits.

The Power of Understanding What Customers Value

Example 1: Costco

Every day, hundreds of thousands of customers in the United States shop at Costco's 500 discount stores. Costco is one of the great success stories of retail in the USA. Why? During our workshops, we've asked who among the attendees shop at Costco. Not only do a great majority of attendees regularly shop at Costco, but they also come from all social and economic categories. When asked why they shop at Costco, attendees typically mention quality products at the lowest price. Then, far behind, come a few additional ones such as a liberal return policy and one-time specials. But what is surprising is what people have to accept to get these benefits. Yet in spite of these nuisances, Costco has 23 millions members who gladly pay the $50 annual membership fee for the privilege of getting quality products at the lowest cost. Why? What does Costco understand that others don't? The short and sweet answer: Costco understands what customers value, which in its case is the guarantee to get quality products at the lowest price. Moreover, Costco understands that its customers are willing to accept many nuisances — many of which Costco's competitors had never before dared to try — to get these benefits. Costco executes on this simple model with total fanaticism.

What does the Costco example have to do with actual products, high tech or not? One key thing actually; Costco understands the trade-offs customers are willing to make. Many product companies survey their customers about their needs in several areas, and they try to match these needs. But Costco is showing us that the right approach is to find the few benefits the customers value highly while delivering the minimum service level they are willing to accept to get these benefits. In other words, they understand trade-offs and priorities. It is because Costco can save a lot of money on these lower level needs that they can deliver these key benefits better than competitors.

Example 2: Google

Larry Page and Sergey Brin had a single idea in mind: "organize the world's information and make it universally accessible and useful." Google flourished because it did a single thing better than anybody else: Search. Do you remember the first time you used Google search engine and found the information you were looking for in the first few links of the first page? Talk about creating value! Google became, and still is, the favorite search engine of the internet crowd. Its application has spread to more and more domains: from places on earth, to scholarly works, and more.

But how do you make money with search engines? You have to create value to advertisers. And Google settled to do a second thing better than anybody else: Online Advertising.

> [Google provides] advertisers with the opportunity to deliver measurable, cost-effective online advertising that is relevant to the information displayed on any given page...Thousands of advertisers use our Google AdWords program to promote their products and services on the web with targeted advertising, and we believe AdWords is the largest program of its kind. In addition, thousands of web site managers take advantage of our Google AdSense program to deliver ads relevant to the content on their sites, improving their ability to generate revenue and enhancing the experience for their users — quoted from Google's web site.

As a result of these two things that Google does extremely well, it generates revenue in the billions of dollars. Not bad for two Stanford graduates who simply understood what people value.

Faced with a Need

Lessons

No mention of successful products would be complete without mentioning the iPod and the iPhone. But before these two there were many others, such as: the Walkman, which the iPod dethroned, the IBM PC, the HP calculator, the Nokia cell phones, The Kleenex, the Polaroid camera to name just a few. All these products were widely successful because they did very few things extremely well at a reasonable cost. It is the lesson we can draw from these examples: Do one or two recognizable things that customers appreciate better than anybody else and you will be on the road to success. Do not try to be better at everything, because you will be priced out of reach of your target customers. Just one, two or maximum three things! That's what it takes. This guidebook will show you how.

Definition

Depending on the situation, we use the terms **BENEFIT**, **POSITIVE**, **PROS**, **GAIN** or **POSITIVE OUTCOMES** to express what the customer thinks or feels he will get from the offering.

Likewise, we use the terms **COST**, **NEGATIVE**, **CONS**, **PAIN**, **SACRIFICES** or **NEGATIVE OUTCOMES** to express what the customer thinks or feels he will have to give.

Value Implies Trade-offs

The Costco and Google examples show that as customers we estimate what we need or want (the benefits) and what we are willing to accept or pay to get it (the costs). This is the fundamental trade-off upon which our evaluation of the value of something is based. Then we compare alternative offerings or possibilities, and the one that has the most value for us is the one we will pick in a free-choice environment. Value is the fundamental criterion behind most decisions we make, be they personal or professional. Our mind is a value computer, constantly evaluating the pros and cons of alternatives.

Attributes of Customer Value

Going back to our Costco example, there are several observations we can make. First of all, some people do not shop at Costco. For example, some people might want a personal service when shopping, or they do not need or want to buy in bulk. For these people, Costco offers inferior value compared to the local store. Value is in the mind of the customer, not in the product. This is a very simple thing to understand, but something that people have difficulty incorporating in their actual behavior. What this means in practice is that you need to define the target customers first to assess value. Your product has no value until a customer contemplates it.

Now suppose it is 10 p.m., and you realize that you forgot to buy the plates for this office party the following day. Again, Costco lost its value for you because it's closed. Value depends on the situation you are in at the time of the need. Many parents of young infants shop at Costco to buy diapers at a great savings. When the child is toilet-trained, Costco has lost some value because value depends also on the objectives of the customer.

Customer Value

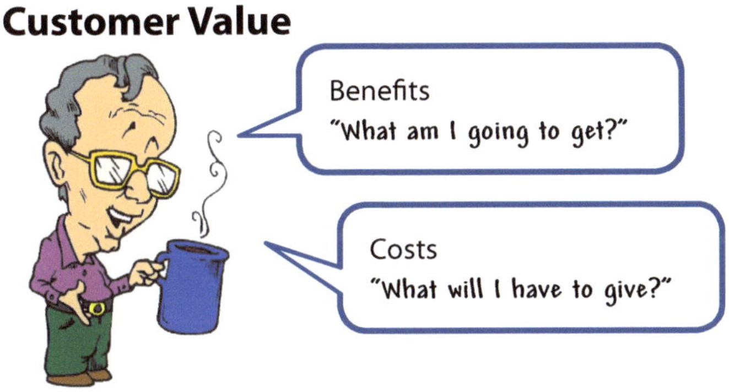

Value is simply the difference between the benefits you get and the costs you have to bear to get these benefits.

Or imagine that you read in a newspaper that, actually, Costco is not the cheapest you can find. Although you had been convinced up until now, from what you heard everybody say, that Costco guaranteed the lowest price, now your trust is questioned, and Costco has lost some value. Value depends on your feeling. Through all these examples, you realize that although the offering never changes, its value is a somewhat volatile quantity that cannot be assessed unless you define a certain number of conditions, all of them involving customers and their decision process.

Faced with a Need

Defining Customer Value

The value of an offering for a particular person is the **PERCEPTION** he or she has of the difference between the **BENEFITS** he or she will derive from the buying and use of this offering under a particular situation, to achieve specific objectives, and the **COSTS** he or she will have to pay to get these benefits. This definition we can summarize with the following formula:

CUSTOMER VALUE = PERCEPTION OF [BENEFITS – COSTS] $_{\text{SITUATION, OBJECTIVES, FEELINGS}}$

Of course, this equation has to be interpreted as a conceptual expression and not as a pure mathematical formula. In particular, not all benefits have the same importance for a particular customer and, therefore, should not simply be added up. We will see in a later section how to take this factor into account.

Essential

"Your product has no value for a particular person until he **BELIEVES** it will allow him to obtain the **RESULTS** he wants in some specific use and situation at **COSTS** he's believes are worth it."

Your product (or competitive products) has no value for a particular person until he **BELIEVES** it will allow him to obtain the **RESULTS** he wants in some specific use and situation at **COSTS** he's believes are worth it.

Please note that in the formula, we have the word "perception." Value is not an objective quantity. Perception means that value can be influenced. The good news is that you, your partners and even your existing customers can increase the value you provide with adequate communication programs aimed toward your future customers. The bad news is that others (your competitors, the media, other customers) can decrease the value you provide the same way. It is important therefore to monitor the value you provide.

If the above definition of value seems too intimidating, there is a working definition that capture the essence of what value is:

CUSTOMER VALUE = IF [WHAT YOU GET] IS WORTH MORE THAN [WHAT YOU GIVE]

Every time we use the term worth, we make a value statement: "Is it worth the cost." "It is worth the time." "It is worth the effort." The "it" is what you get. Although "it" is often left unarticulated in the common expression, the good marketers probe to understand and articulate what the "it" is made of.

The Number One New Product Success Factor

Robert G. Cooper, in his book Winning at New Products, has characterized the most important success factor for new products as "A unique, superior and differentiated product that offers superior value to the customer." The goal of this guidebook is nothing less than helping you achieve this success factor. Don't all companies want to do that? Then why are there so many so-so products in the market? The short answer is that many companies are too product-centric and do not understand 1) what their customers value and 2) what differentiation is and why it is important. By the end of this chapter, you will be able to understand these two concepts. By the end of this guidebook, you will be able to execute on this understanding.

Ideas for Actions

"It" is Made of Value

Actually, "it" is what **CREATES** value. Therefore, you must identify a strong need and the higher in the organization the person who has the need is, the higher the value you can provide, and the greater (or larger) the business you can generate. Is your offering solving a technical problem or is it increasing profit? The value is not the same in both cases, and the second has definitely more value. Learn to express your benefits as addressing specific business improvement issues such as: reducing cost, time to market, resources, waste, or increasing revenue, profit, productivity, throughput, quality, satisfaction, competitive advantage, etc. And if your offering cannot specifically and convincingly address one of these, start immediately to redefine it. Otherwise, beware the competitor that can.

Essential: "The successful product offers a unique, superior and differentiated product that offers superior value to the customer."

Key Points

- The number one success factor for new products is delivering superior customer value and being well-differentiated.
- Value is the only criterion people use to make a purchasing decision (assuming free choice, of course).
- Value is NOT in the product or features.
- Value IS in the customer's mind as benefits, and each customer has a different appreciation of the value he or she will derive from something.
- Thinking of value requires you to think of what you and your company do from the customer's point of view.
- Value depends on the situation, the usage, the objectives and the preconceived ideas of a customer.

Thinking Like Your Customers

Important: A firm must develop a core competency in understanding what customers value and in transforming this knowledge into whole products that create superior value.

Thinking About Value Requires Thinking Like a Customer

Since value is in the mind of the customers, marketers need to become experts at reading customers' minds. First, you need to think like your customers. It's not easy. You spend most of the day working to make your products and your company more successful. Your customers do not. They spend most of their days working to make *their* products and *their* company more successful. And then one day, you visit them, and because you spend two hours with them discussing your product, you think that they spend the rest of the year thinking about your product too. What else would they be thinking about? If you want to create value for your customers, you need to know what else they are thinking about. We will examine in the next section how best to do it.

Important: Knowing your customers require knowing what results they want to achieve to meet their goals and objectives.

What Does it Mean to Know Your Customers

We often ask workshop attendees, "What does it mean to know somebody?" It's interesting to hear the variety of answers: for example, "knowing their occupation, their activities, their reputation," or "knowing their needs," or "knowing their character, or what they like and don't like." Even if these answers are technically correct, they are also incomplete. We then follow with a second question: "Do you know your spouse?" After the usual laugh, people in general say "yes." The next question then is, "Why do you say you know him or her?" That question makes people really think. Spend a moment thinking about your answer before continuing…

Knowing somebody means being able to anticipate his or her behavior, attitude or action in response to any situation the person is confronted with. This definition applies equally well to your customers, your competitors, your colleagues or your spouse.

Now think a minute. Pick a customer or a customer segment. Do you know that person or segment according to this definition? Please note that I have not mentioned your product or your company, because so far your product and your company are irrelevant. How can you anticipate the reactions of a person or a company? That's what we are going to examine next.

Faced with a Need

Why Customers Buy: The Compelling Reason to Act (CRTA)

Search for the Customers' Compelling Reason to Act

It's pretty common to say that marketers, salespeople and the company must understand customer needs and develop products that address these needs. But it's also very common for anybody in sales or marketing to think they've identified some good customer needs and find later on that the customers don't buy the product the company has developed to address those needs. Why so? Because of two reasons: 1) Everybody has more needs than can be satisfied and 2) Marketers and salespeople usually express needs in term of product features. But needs may change or disappear, or be met by another product. Remember what we said earlier, that value depends on many things, and it changes over time and circumstances. Only a few needs are really critical. These are the ones you must learn to identify and focus on. We are calling these needs **COMPELLING REASONS TO ACT (CRTAs)**. These are the needs that the customer MUST address now; hence these needs are "compelling."

The Google example mentioned earlier is a good example of this difference between needs and CRTAs. Now in the Google case, there are two customers: the end-user and the advertiser. Google had zeroed in on the CRTA of each. For the end-user, finding the right information in the minimum time is a very convincing CRTA. For the advertiser, finding more leads or prospects at the minimum cost is precisely what advertising is all about. And because Google executes on these two CRTA equally well, it has risen into an unavoidable resource.

Think about a product you have. Can you define the CRTA it is addressing and what you are actually doing to constantly improve its performance in this area? STOP! Remember, "No Feature."

How to Know Your Customers', Your Competitors' and Pretty Much Everybody's CRTA

People and companies act because they have goals and objectives they want or need to reach. Nearly all companies have annual and three-year plans that outline a few key goals and objectives that management commits to achieve. These goals and objectives are then rolled down the organization to become division, business-unit, department and individual goals and objectives, and budgets and resources are allocated. Obviously, there are obstacles, opportunities, threats and issues on the road to achieving these goals and objectives. Removing these obstacles and addressing these opportunities are the relevant CRTAs until the goals and objectives are met. Helping your customers articulate and meet them in an effective way provides significant value. So what is critical to identify is not their "needs," but rather what results or outcomes your customers want to achieve, and under what conditions of time, money, etc. If you do not know the goals and objectives of your customers, what's in their way to reach them, and what results they want to achieve, you cannot say you know your customers and you cannot create value for them. In other words, knowing your customers requires knowing what results they want to achieve to meet their goals and objectives and how they want to achieve them.

Probing Questions for Identifying and Qualifying the Customer's CRTA

Problems

First, you must identify what problems the customer is facing. The customer may be aware of these problems, but if not, you will need to ask pointed questions. Here are a few examples:

- "What keeps you awake at night?"
- "What are the top three objectives you must achieve? Among these three, which one is the most important? Why is it the most important?"
- What is the biggest challenge you are confronting right now? What is the number one obstacle you are facing right now?

Definition

Compelling Reason to Act (CRTA)

A clear problem or opportunity that can preempt or help a company or an individual to achieve critical goals and objectives.

Faced with a Need

Once you have identified the problems, you will want to understand how serious they are. Consequences, payback and urgency are then the areas to probe next.

Consequences

What would be the consequences for the individual and the company if nothing is done?

- "Why is that objective so important to you?"
- "What would happen if you do not reach this objective?"
- "What would be the consequences of doing nothing?"

Payback

Payback is the return they expect from getting the expected outcomes:

- "How much time would you gain if you implement this program?"
- "How many people would that free up?"
- "How much space could you free up with such a change?"
- "How much money would that save?"

Urgency

How critical is it to address this CRTA now?

- "When would you need to reach this goal? Tell me why this is important."
- "When could this problem become a big problem?"
- "When would you like to see the first positive results?"

Beliefs, Values, Policies and Assumptions

Faced with the same goals and objectives and the results they want to achieve, different people and companies will react differently and make different choices based on beliefs, values, policies and assumptions they have about a variety of subjects, including themselves, their company, their customers, their industry, your company and your competitors. A company may, for example, want to be viewed as friendly to the environment, which would shape the kind of choices they might make. Or they may want to be perceived a strong leader, which would also shape their decisions. Probing questions to identify beliefs, values, etc.:

- "Tell me how the policies of your organization could help or hinder this program."
- "Tell me the role that the company values could play in that decision."
- "Give me an example when the company values, beliefs or policies interfered with an initiative like this."

Capabilities

The strengths and weaknesses of the organization will shape the ways your customers view value and solutions to their problems. Does the problem they face fall into a domain of strength or weakness? If it is in a weakness area, chances are that they will search for outside help in some form or another. On the other hand, if their problem falls in an area of strength there is little chance for your company to add significant value. Probing questions to identify capabilities:

- "What kind of internal and external resources do you believe you will need to solve this problem?"
- "How did you address such a problem in the past?"
- "What outside resources do you see needed for this project?"

Faced with a Need

Customer Value Drivers

What Customers Buy

Faced with a CRTA that makes them looking for a solution, customers define a set of criteria by which they will evaluate their options. These criteria could be well-articulated or fuzzy and ill-expressed. Your role as a marketer is to understand these value drivers very well. If customers have difficulty in articulating them, your role as a marketer is to help customers articulate these criteria — but even then, it will only include the conscious needs. In most circumstances, customers have unexpressed needs. Observing the customer in his environment is a proven technique to identify these unexpressed needs.

Definition

Customer Value Drivers (CVDs)

The criteria that customers use to evaluate competitive offerings in their ability to satisfy their Compelling Reason to Act (CRTA).

Customer Value Drivers (CVDs) are the criteria, expressed as benefits, that customers use to evaluate competitive offerings in their ability to satisfy their own CRTAs. These CVDs should be identified in all categories of benefits and costs that the customer considers (more on this in the following section). Remember that the customer is not always conscious of the criteria he will use to make a decision, particularly the emotional influences; furthermore, many of these CVDs can be deep in the customer's psyche and, if not properly identified, can cause a company to act upon the wrong reasons. In addition, the CVDs are not all of the same importance; consequently, you must probe for hints in the various categories. We often meet clients who say that they are not satisfied with their market position. In many cases, the reason may be traced to a lack of understanding important CVDs. As marketers and salespersons, we are all focused on thinking "product" and "solution" first. But this is only one of the areas that customers consider. Remember, the objective for the customer is to solve the problem or address the opportunity generating his CRTA. The objective of the customer is not simply "I want to buy a product."

Creating Customer Value

More than Just Technical Benefits

Faced with articulating customer requirements for a new product in their product line, most product teams develop a list of technical areas to probe. That process usually leads to better, faster, lighter, cheaper, etc., products. However, the hot new technical advantage soon becomes passé as the competition catches up. Soon it is time to develop another round of better, faster, lighter, cheaper, etc., products. This constant pursuit to maintain the technical advantage over that of your competitors' — this "arms race" — can become quite costly; moreover, it runs the danger of overshooting the real needs of the customer, which opens the door for your competitors to find innovative ways to address the customer's total needs.

Important

Value is created when you can provide a solution to a customer's CRTA that will allow him or her to obtain the desired outcomes at an acceptable cost.

When entering a new market, it is critical to understand what opportunities exist to add more value beyond what is already being offered. For example, the areas in which value could be better created than by existing competitors are areas in which opportunities exist for you. And these areas are not only technical. It is to help the marketer probe the these areas that we have developed the following model of value.

Over the years, we have found that CRTAs and value drivers can be found in one to five categories of benefits and one to five categories of costs.

Benefit Categories

1. Technical
2. Operational
3. Economic
4. Emotional
5. Network

Cost Categories

1. Monetary
2. Resources
3. Troubles
4. Risks
5. Lost Opportunities

These lists of the five dimensions (or categories) of Benefits and the five dimensions of Cost are further explained in this guidebook. Benefits, like costs, should be probed during customer interviews to avoid falling into the trap of focusing only on the "banner specs."

Creating Value by Increasing Benefits

Benefit Category #1: Technical

When the customer has a technical problem that preempts him or her reaching an objective, the CRTA can be expressed as "I need to solve this technical problem; otherwise I will miss my objective, and this is not acceptable." For example, an R&D department has identified a major glitch in the prototype of a new product that puts in jeopardy its time-to-market objective; none of the tools available in the department or the company can address that problem, because it comes from a new kind of IC's and no tool can go fast enough to detect that glitch. The R&D department has a technical CRTA that might be addressed by a technical offering such as a new-generation tool, allowing the customer to derive technical benefits.

Increasing technical benefits is the first thing most marketers think about — for good reason. But improving technical performance is just one category of benefits to consider. There are four other categories that are worth exploring.

Ideas for Action: What to Probe for

When trying to understand the technical benefits that you can deliver to customers it is always recommended that you look to the total technical solution including usability and usefulness. A product is rarely isolated and is often part of a larger system or context. What is the bloc diagram? The process flow? What issues does the customer face upstream and downstream? What results does the customer want to achieve? What is on her way? Is the result your product provide the final one? Or an intermediate step (such as an image, a picture, a number)? Very often, you will find that there are opportunities in exploring the total system/solution.

Benefit Category #2: Operational

When the customer has an operational problem that preempts him or her reaching an objective, the CRTA can be expressed as "I need to solve this operational problem; otherwise I will miss my objective, and this is not acceptable." Operational CRTAs cover a wide spectrum. For example, a company has an objective to improve quality. They recognize (maybe with your help) that they are weak in documenting processes. This company

Faced with a Need

has an operational CRTA that might be addressed by a service-oriented offering such as training and consulting services, allowing the customer to derive operational benefits (improving quality).

Ideas for Action: What to Probe for

Look beyond the immediate function of your product and look for the critical process it is part of, who is the "customer" of this process, and what the outcome as seen by this person is. For example, an oscilloscope can be seen as just an instrument to detect bugs, which it is. But it can also be seen with its operator, as a critical part of a lean design process initiative, for example. The later is of higher potential value than the former, because it addresses a higher level business problem. The key here is to go beyond the immediate usage of the product and explore business benefits. How can you help the customer be more efficient?

Benefit Category #3: Economic

When the customer has a financial objective such as ROI, sales or cost, the CRTA can be expressed as "I need to evaluate all the decisions and investments we make in light of how they contribute to reaching our financial objective; otherwise we will miss them, and this is not acceptable." For example, a company has an objective of increasing its ROA (Return on Assets) by 20 percent over three years. This company has an economic (or financial) CRTA — Return on Assets being expressed as Financial Return divided by Assets. An offering that can get more return out of existing assets or can get a comparable return with fewer assets will provide value. These are economic offerings, and they allow the customer to derive economic or financial benefits. In business, all buying decisions involve some kind of economic return. But for an offering to be classified as an economic offering, the economic return it provides must be measurable and quantifiable.

Ideas for Action: What to Probe for

Operational benefits often translate to economic benefits that can be quantified. For example, many large purchases are made on that basis and when the financial benefits are taken into account, the customer can calculate the return on investment (ROI).

Benefit Category #4: Emotional

This is when the customer has objectives that are beyond the material world; this category addresses the domain of feelings, emotions, spirituality and other intangibles. Consumers are driven by emotional CRTAs and value drivers; even if they are not conscious of it, they buy goods for reasons that go well beyond their pure

Benefit Categories

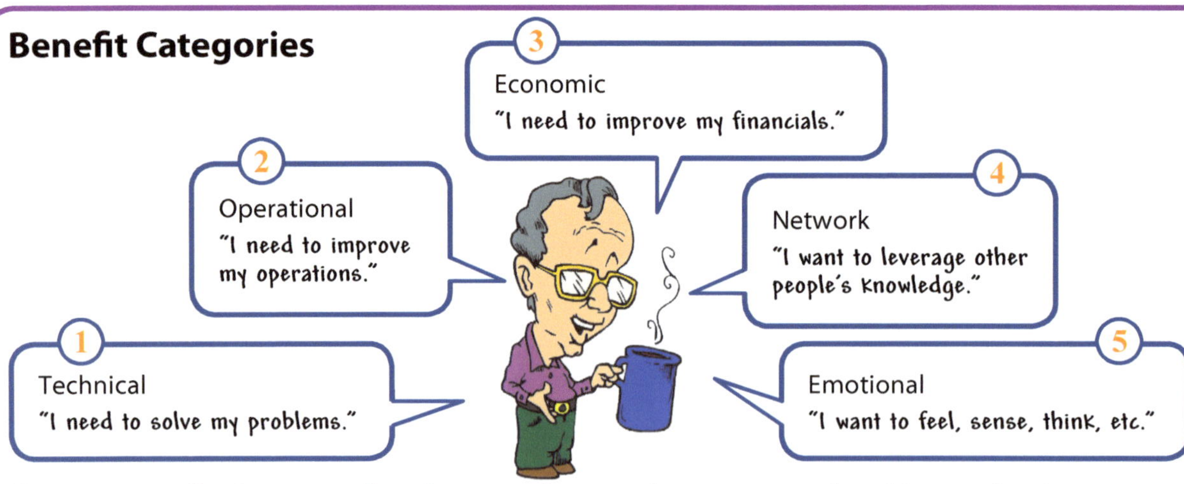

Customers use five important Benefit categories to evaluate your product. Many technology companies rely on Technical Performance (#1). While that category is important, there are at least four other categories, and these are all opportunities by which to differentiate your offering. When you interview your customers, remember to probe beyond technical or "banner specs."

utility. For example, a young lady wants a Vuitton bag and a Chanel perfume to feel better, or a middle-age man wants a sports car to feel more powerful.

There are two ways to create emotional benefits: 1) create a pleasurable experience with the product at all levels, or 2) create a pleasurable experience interacting with the company. In the first category, you see companies now putting more and more emphasis on designing desirable and pleasurable products that create a strong bond with their customers and users (iPod anyone?). In the second category, companies are also putting emphasis on developing "moments of delight" at the critical time when customers interact with the company, such as a customer support call. We call these "emotional offerings", and they allow the customer to derive emotional benefits.

Companies also have emotional CRTAs. For example, increasing the retention of valuable employees could be a company objective, which would generate emotional benefits because employee retention fosters within the workforce feelings of being valued, appreciated and able to express one's potential.

Luxury goods, theme restaurants and entertainment are examples of emotional offerings for consumers. Enterprise consulting services, branding consulting services and the like are business emotional offerings. All companies and products generate feelings when we interact with them, and these feelings can be pleasant (we want to repeat them), neutral or unpleasant (we want to avoid them). A company that has perfectly understood this category of benefits is Apple. Everything Apple designs is to generate feelings. Does it apply to all industrial products? Yes. The user of the product might like or dislike the user interface; the buyer might like or dislike the people interacting with him.

Customers are people with feelings, and these feelings play an important role in how your offering ranks among the alternatives. Among all the things you want to know from a customer, emotional needs are the ones that you cannot openly discuss. It takes empathy to understand the emotional needs of customers. Asking the right questions, creating an atmosphere of trust and interpreting between the lines are skills you will need to master. Because emotional benefits generate loyalty, developing emotional benefits has recently taken center stage in product management.

Ideas for Action: What to Probe for

Designing a pleasurable customer experience is of critical importance if the user of the product has a word to say in the choice of products and vendors. The human-machine interface has then become a key source of effort for many companies. User labs are the best way to observe customers using the product. Another way is to shadow a customer on her site (office or home) and observe what she goes through

Brand attraction and loyalty is also mainly based on emotions, and it is critical that companies assess the power of their brands to attract and retain customers. One of the key feelings that business companies live on is "trust." Trust generates loyalty. How do you create, develop and maintain trust?

Benefit Category #5: Network

We live in an interconnected world. Our efficiency and effectiveness can be highly improved by tapping into the knowledge and capabilities of others, be they inside or outside the company. We need, therefore, to have easy access to information. When we ask attendees to our seminars who uses Microsoft Office, everybody raises a hand. When asked why, the answers all converge toward this reason: "because everybody else uses it, and I need to communicate easily with others." Is it the best product? That question is not so important. What is important is to be compatible with others and to be able to leverage other people's knowledge and services. The Internet and wireless communication explosion have suddenly brought this potential benefit to the forefront. We call these offerings network offerings, and they allow customers and products to connect to other people and products and improve their performances. Web sites such as eBay, MySpace and Facebook are examples of pure Network offerings. Companies that can include some Network benefits in their offerings can gain considerable advantage, because as the

Important! With the wide availability of the Internet, network benefits, or the ability to tap into other knowledge sources, have become one of the most powerful ways to add value.

Faced with a Need

success of these companies exemplifies, this kind of offering tends to become "winner-take-all" in their markets. Microsoft and Amazon, for example, continually add network capabilities to their basic offering with great success.

Network benefits are not to be confused with the usual relationship between people, in particular between the vendor personnel and the customer. Relationships create emotional benefits such as trust or peace of mind. Network benefits relate to the customer's ability to better achieve goals by tapping into the resources of a network of other products, people or knowledge.

Ideas for Action: What to Probe for

The world is evolving into a set of networks. What network is the customer part of, and how can you contribute to developing this network. More and more customers rely on these internet networks for their personal and professional development. If you have or can have many users, you need to consider organizing them in a network. What would make the customer more effective if he could network with others? Many tech products have a user group. How do you help people in this group to network?

Creating Value by Reducing Cost

Remember that when making a decision, customers look at more than the benefits they can get. They also consider the costs they will have to accept to get these benefits. It is important to note that reducing any of these costs becomes a source of benefits. Costs that customers consider also fall in five categories.

Cost Category #1: Monetary

Acquisition cost, of course, is an important cost. But the price of the offering is just one of the costs customers experience. To understand the cost from the customer's perspective, you have to consider cost as Life Cycle Cost (LCC) or Total Cost of Ownership (TCO).

Ideas for Action: What to Probe for

All products acquired by a company go through the same phases: selection, procurement, receiving, inventorying, dispatching, paying, training, operating, supporting, maintaining, managing assets, upgrading, and disposing. All these tasks cost money; sometimes much more than the initial purchase. For example

Cost Categories

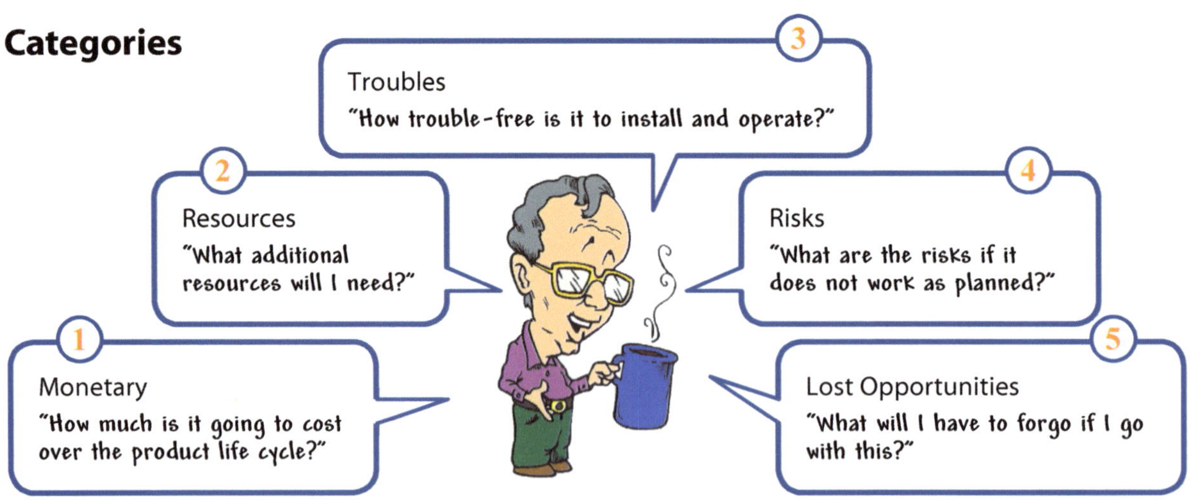

Customers use Cost categories to evaluate your product. There are at least five opportunities to show that your product may reduces cost. When interviewing customers remember to probe for Cost Reduction.

the Total Cost of Ownership (TCO) of PC's in a firm is several time the initial purchase price. Companies that can provide significant savings on this lifecycle cost do have an advantage. Do you know the Total Cost of Ownership (TCO) of your product and do you have an active program to decrease it? If you don't, you are at a disadvantage compared to a competitor that can articulate it.

Cost Category #2: Resources

At all stages of the life of the offering, resources other than money will be needed. Resources are people, time, capital, square footage, etc. Time is a particularly important one, as most companies are driven by reducing the time it takes to do things. Most of these resources can also be expressed in monetary terms, but it is good for marketers to keep them separate to ensure that they are well-identified.

Ideas for Action: What to Probe for

All these areas need to be investigated — in particular, people and time. Questions like what would be the benefit for you if you could accomplish the task in half the time, or with half the number of people? Actually, more and more diagnostic products specify their time to results. This is an example of focusing on the customer outcome rather than on the technical need itself. Focusing on time to results forces to investigate all the steps from beginning to end and how to decrease or eliminate the time wastes.

Cost Category #3: Troubles

Troubles are psychological costs. Troubles happen all along the life of any offering, when something does not work as planned. They cannot be quantified, but they can impact the efficiency and effectiveness of a person or a group. Trouble is a cost, but removing troubles and making the life and work of customers easier creates value.

Ideas for Action: What to Probe for

Identifying the sources of pains and troubles and looking to eliminate them can unveil powerful sources of value. Questions like "What is bothering you when you do XXX?" or "What difficulties do you encounter when accomplishing that task?" Can lead to opportunities. Most often than not people have learned to leave with these pains, so they have been relegated to the unconscious; observations and other ethnographic enquiry approaches need to be used. The infamous blinking 12:00 VCR of yesteryear is a prime example, but how many alarm clocks are a pain to use?

Cost Category #4: Risks

And what if it does not work as planned? Customers consider this as an emotional cost.

Cost Category #5: Lost Opportunities

Buying a particular offering means not buying another one, and therefore not being able to do something else. This is a lost opportunity that customers often consider when making a decision.

Migration of Customer Value over the Technology Adoption Life Cycle

Is one way to create value for customers better than another? The short answer is yes. Of course the customer is always the one who defines value. But there is a general principle that can guide marketers to explore in more depth certain dimensions of value over others. This principle is quite simple: As a new product category moves from the early adopter phase to late majority, value moves from mostly technical to mostly emotional.

When a new product category appears, the first to try it (the early adopters) are willing to accept risks to see how they can use the new capabilities. Understanding these new capabilities and what users need to do to put the technical aspect of the offering to work for them is their main concern. At this time, the product is mostly incomplete, imperfectly addressing customer problems. But for the benefit of having access to the new

Faced with a Need

possibilities, early adopters accept some risks and some extra work, including completing the product to meet their needs.

As improvements of this new product category are introduced, another group of customers becomes interested. These customers now are interested in how this product will solve problems they have, while earlier it was to investigate what they could do with the new product. These customers are still willing to take some risks and actually build a part of the solution themselves to get an advantage. At this stage, operational benefits (the ability to help customers develop and use a solution) grow in importance.

As applications develop in the market and products evolve to address a few of them, more and more customers are asking for a complete solution, and proof that it will actually deliver on the promise. At this stage, economic benefits grow in importance.

After this stage, the product category evolves in a few well-defined, cataloged solutions, and the customers become more and more knowledgeable about the use of the product. Technical and operational issues have been addressed, and the market has settled on a few key competitors that now fight for market share. With technical improvements not able to deliver much additional value, competition evolves toward products of lower cost and greater usability and desirability, such as ease of use, pleasing aesthetics and image. At this stage, it is emotional value that grows in importance. Network benefits also represent a growing opportunity as the product moves from early adopters to late majority as the number of users increases (if the category is not itself a purely relational product from the start).

The relative importance of costs, the other term of the value equation, also changes along the life cycle. At the introduction of a new category, purchase price and resources are the key costs people consider. But as the category matures, the other components of the cost of ownership become more important, and the customer becomes less willing to accept risks and troubles. Understanding this evolution — in particular the opportunity that the change in expectations from one category of buyers to the next represents — has been instrumental in many gains and losses in market share and leadership.

Important! As a new product category moves from the early adopter phase to late majority, value moves from mostly technical to mostly emotional.

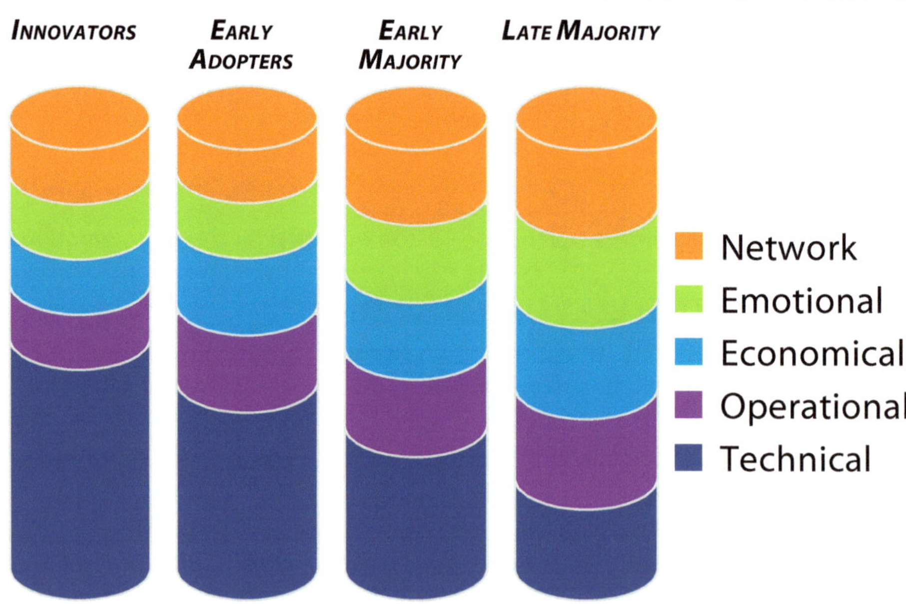

Products move through a well-documented curve called the Product Life Cycle, similar to the Technology Life Cycle. At the earliest stages of the cycle, the focus is on performance (banner specs). As your product matures and competition intensifies, additional categories factor into the decision to purchase. Note that towards the end of the cycle, Technical Performance loses value.

Dimensions of Value

Representing and Quantifying Value

Formally and informally, consciously and unconsciously, customers evaluate the value of an offering by assessing its benefits and costs. To really understand value, you need to model this evaluation process so that you can reliably provide the same answer as the customer's.

First, identify and express the most important CVDs of your target segment. The objective of customer visits — which this guidebook will help you plan and execute — is to identify these CVDs and rank them in importance.

To assist you in identifying these CVDs, we have listed common benefits and costs for each of the five categories discussed earlier. This list is applicable to a wide variety of situations, but you should customize it to fit your own situation and, most importantly, to reflect the way the customer would express them.

Benefits

Dimensions of Value	Typical Examples
1. Technical	• Ability to address existing and anticipated technical problems. • Ease of learning and operating the product over time. • Ability to provide solutions to complete existing and anticipated problems. • Ease of learning and operating the system over time.
2. Operational	• Ease of maintaining the product through its life. • Ability to reduce the time it takes to perform a task. • Ease of integration into company processes and operations. • Ability to improve company processes.
3. Economic	• Ability to increase revenue. • Ability to decrease cost.
4. Emotional	• Positive experience with the product. • Positive experience with the services of the company. • Positive experience with the employees of the company.
5. Network	• Ability to improve performance by connecting with other products. • Ability to improve performance by connecting with knowledge bases. • Ability to improve performance by networking with other people.

Faced with a Need

Costs

Dimensions of Value	Typical Examples
1. Monetary	• Purchase price. • Total cost of ownership.
2. Resources	• Amount of capital/space needed. • People and time needed. • Information needed.
3. Troubles	• What additional things "I" (the customer) will have to worry about as a result of having this piece of equipment.
4. Risks	• Impact on the organization if something goes wrong. • Impact on me if something goes wrong. • Reward for me if everything goes right. • Impact on the environment.
5. Lost Opportunities	• What will I have to forgo if I purchase this piece of equipment?

Qualifying Value: Product Value Profile™

Differentiation

Differentiation is another term widely used in marketing circles, but why is it important? With superior customer value, differentiation has been characterized as the number one success factor for new products (Robert G. Cooper, Winning at New Products, (3rd ed.), Basic Books, 2001). Differentiation is the foundation on which positioning is built. And positioning is what your customers will "store" in their minds about you and your product.

Al Ries and Jack Trout, in their 1981 landmark book Positioning: the Battle for your Mind, use a simple but powerful metaphor to explain why customers prefer some companies and products to others — they consider the customer's mind to store many little "boxes" labeled "best buy for this product category." Each box is a category and can only contain a few products/companies. Which products/companies will win the right to be in the box, knowing that each box cannot host more than a few names? For example in the laptop box, it is likely that HP, Dell, Apple and the Thinkpad are prominently there. Maybe there is a fifth name, but that's it.

Important: Well-differentiated products offer better solutions to the most important customer value drivers.

Important: For a product to be well differentiated, it needs to be **MEANINGFULLY DIFFERENT** from the other products in the same category.

See Also: **Product Value Profile Example**

If you want to get in the box, you have to dislodge the existing occupants — not an easy task, although it can be done. It's much easier to try to find boxes that are not yet full so that you can gain the right to be in that box with a select few others. And when you are in a box, it's likely you will stay there unless you make mistakes that the customer cannot accept. Finding or creating these boxes and filling them with your product is the role of positioning. And positioning builds on differentiation. If you want to get in a particular box, you have to be different from whatever is already there. It's easy to be different, you say. Yes, but it's not enough to just be different. You need to be "meaningfully different." Being meaningfully different means being better in satisfying the value drivers that customers value most. We have come full circle: The most important success factor requires a deep understanding of the customer value drivers. There is no other secret besides the ability to execute on this understanding.

Important: If you have the print version of this guidebook, e-mail the authors for a copy of the Microsoft Excel file: voc@rapidinnovation.com.

Represent and Evaluate Products Side-By-Side Graphically

Compare the Value of Your Product with that of Your Competitors'

A Product Value Profile graphically compares your product with your competitors' products, and the ability of each to meet the most important CVDs, showing clearly which product is best.

For example, if the comparison shows that a competitor's product is better than your product in two of the three most important CVDs — even if your product triumphs in all but one of the remaining CVDs — it is an indication that your product is weak, undifferentiated or mis-targeted, that it tries to be good across the whole set of CVDs without being excellent in the most important ones.

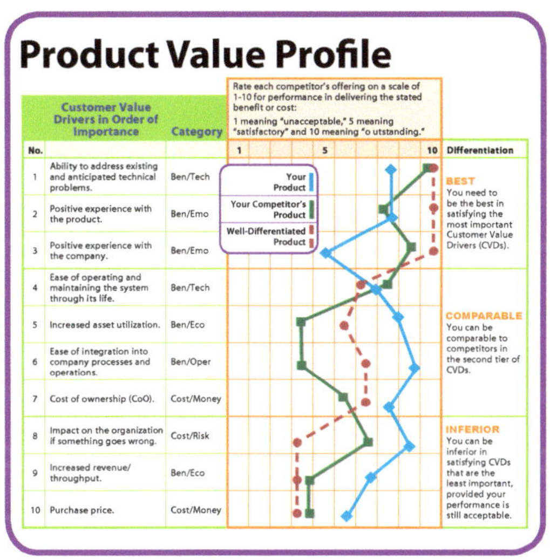

Compare the Value of All Products with that of a Well-Differentiated Product

To further illustrate this point, a third line is used to represent an ideal well-differentiated product. A well-differentiated product is one that

Faced with a Need

delivers the best performance in the most important CVDs of the targeted market segment. A product that is the best at the top three CVDs has all chances to be a winner in the broad market. For example, Toyota is widely recognized as leading in the following three CVDs: reliability, mileage and life-cycle cost.

A product that is the best at a single top CVDs has all chances to be a leader in a niche market; for example, Volvo has been for a long time widely associated with safety — a top CVD. How many top CVDs you target for being the best depends on your marketing strategy. The important point here is that you don't need to be the best at everything. One, two or three CVDs is all you need to carve out a strong position in your target market. The fewer CVDs you are recognized as being the best at, the better your product is differentiated, but your market may be smaller.

For example, Costco has identified low-cost and quality, as its top two CVDs. Audi, a German car manufacturer, focusses on design, performance, technology and safety. Apple focusses on design (emotional appeal), innovative products and ease of use. If you want your product to develop a strong position, you need to be the best at these top value drivers, or find a segment in which you can be the best at the top CVDs.

The next tier of value drivers are the ones for which you can be equal to the competition, and the last tier are the ones that are least important and for which you can afford to be inferior to the competition, providing that your performance does not become a major objection.

Products that try to be good at everything seldom reach market leadership. Exceptions exist of course, and they are the stuff of legends. Take Lexus for example — it took the luxury market by storm and indeed was the best at everything, even delivering new benefits, like customer service, that had been neglected by its competitors But perfection is seldom achievable. The pursuit of perfection can be like Don Quixote tilting at windmills. But the opportunity is there for the product manager to reallocate the effort placed on the least important value drivers and develop a Product Value Profile that maps closely to that of the well-differentiated product.

As an example of such an approach, General Motors, once the worldwide leader of car manufacturers because it understood the American consumer value drivers better than others, suddenly started to lose market share following the first oil crisis in the early 1970s. That structural change in the environment triggered a change in the American customer value drivers. Suddenly, gas mileage became a top CVD. GM's products' value declined, as it was unable to deliver high-gas-mileage cars to this growing segment. Then came the quality revolution, and quality became a top CVD. GM's products' value declined even more as it was year after year ranked below the likes of Toyota, Honda and the other Japanese manufacturers. Then came design, and the same story repeated. The rest is known: Chapter 11. GM finally reacted, and it has since driven its product development to place emphasis on design, safety, quality and gas mileage. Now attractive designs have started to roll out of GM chains again. Many of its cars are now ranked on top of the charts for initial quality, ahead of Toyota! Will GM introduce cars that cause people to once again flood the dealership?

Level 3 Detailed Process: Charting the Product Value Profile

See Also: Product Value Profile Example

1

List the Most Important CVDs

Enter in column B the customer value drivers in decreasing order of importance, as the customer would express them.

We have limited the tool to the 10 most important ones, as that is usually enough to characterize the customer's decision process.

Important

The fact that there are 10 categories forces you to look beyond technical CVDs — a rare pursuit in technology companies. But by looking at your market and product from a wider angle, you will discover value opportunities ignored by others.

Important

Make sure you have properly identified both the benefits and costs.

The list provided in the previous section is a good start, but you need to make it fit your market and what your customers would say are most important.

Be sure to articulate less obvious benefits and costs — in particular, emotional benefits.

2

Enter Your Customer's Ratings

Enter in column C the benefit or cost category of each value driver.

If you only find technical benefits, think again.

This step will help remind you to explore value drivers beyond the obvious technical ones.

Important

Identifying CVDs requires a good knowledge of your customers, knowledge that is typically acquired by a combination of qualitative and quantitative investigations — activities that will be explained in the following chapters of this guidebook.

3

Rate and Plot Product's Ability to Satisfy

Rate the product's ability to satisfy each value driver, using a 1 to 10 scale.

Enter a distinctive dot at the corresponding location in column D.

The Product Value Profile is the broken line joining these dots.

4

Compare Competitors' Offerings

Repeat Step 3 for the other products you want to profile.

Important

Critically observing your Product Value Profile will help you assess how strong and differentiated your product is.

Your Product Value Profile should resemble that of the well-differentiated Product Value Profile.

Continued

Faced with a Need

CONT'D

5

DETERMINE THREE TIERS OF PRIORITY

Now that the Customer Value Drivers are listed in order of importance, divide the items into three tiers.

You need to be the **BEST** in satisfying the most important CVDs.

You can be **COMPARABLE** to competitors in the second tier of CVDs.

You can be **INFERIOR** in satisfying CVDs that are the least important, provided your performance is still acceptable.

6

RANK THE TOP TIER

For the top tier of value drivers (three, in our example), choose three levels that are superior to the best-performing competitor products on at least two value drivers; you can of course target all three if it makes good business sense.

7

RANK THE MIDDLE TIER

Select the average value of the competitor products, giving higher rank to the more important competitors.

8

RANK THE BOTTOM TIER

Select values that are one to two points inferior to the average of the competitor's product, making sure, however that you do not fall into the unacceptable category.

Important

The key objective of this guidebook is to equip you with the skills to identify critical CVDs.

Example: Product Value Profile

Rate each offering on a scale of 1-10 for performance in delivering the stated benefit or cost: 1 meaning "unacceptable," 5 meaning "satisfactory" and 10 meaning "outstanding." (D)

No. (A)	CVDs in Order of Importance (B)	Category (C)	Differentiation
1	Ability to address existing and anticipated technical problems.	Ben/Tech	**BEST** You need to be the best in satisfying the most important Customer Value Drivers (CVDs).
2	Positive experience with the product.	Ben/Emo	
3	Positive experience with the company.	Ben/Emo	
4	Ease of operating and maintaining the system through its life.	Ben/Tech	**COMPARABLE** You can be comparable to competitors in the second tier of CVDs.
5	Increased asset utilization.	Ben/Eco	
6	Ease of integration into company processes and operations.	Ben/Oper	
7	Cost of ownership (CoO).	Cost/Money	
8	Impact on the organization if something goes wrong.	Cost/Risk	**INFERIOR** You can be inferior in satisfying CVDs that are the least important, provided your performance is still acceptable.
9	Increased revenue/ throughput.	Ben/Eco	
10	Purchase price.	Cost/Money	

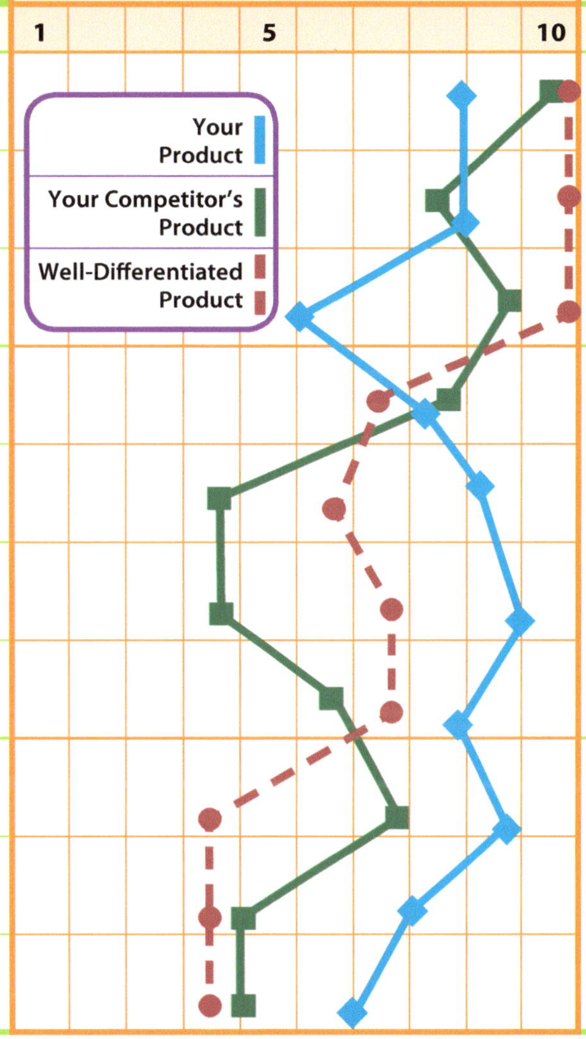

A Product Value Profile graphically represents the ability of competing products to meet the most important CVDs (Customer Value Drivers) for a target segment. The well-differentiated product is one that satisfies the most important CVDs better than any competitor, while being comparable to them on the next tier and even inferior on the last tier.

The comparison here shows that Your Competitor's Product is better than Your Product in two of the three most important listed CVDs — even though Your Product is better than Your Competitor's Product in all but one (CVD #4) of the next tiers — this is an indication of a weak, undifferentiated or mis-targeted product, one that tries to be good across the whole set of CVDs, without being excellent in the most important ones.

In this example, Your Competitor's Product has all chances to be a winner in the broad market. Products that try to be good at everything seldom reach market leadership, or they leave money on the table. But the opportunity is for the product manager of such products to reallocate the effort placed on the least important CVDs to develop a Product Value Profile close to the dotted red line in the figure.

Faced with a Need

Quantifying Value: Product Value Index™

Graphically representing the value of competitive products is very useful to rapidly appreciate the strengths, weaknesses and differentiation of these products. But it does not answer a fundamental question: Which product has more value for the target segment? If you can answer this question, by definition you can assess who will gain market share. To answer this question, we will resort to an extension of the previous tool: the Product Value Index (PVI).

> Essential
> "A Product Value Index assigns a single number to the value of a product for a given market segment or customer."

A PVI assigns a single number to the value of a product for a given market segment or customer; the higher the number, the higher the value and logically the stronger the market position could be. The Product Value Index is based on an extension of the definition of value; it goes one level beyond the basic definition we gave at the beginning of this chapter. Remember that we defined value as benefits minus costs, where benefits represent the ability to satisfy customer value drivers. But not all benefits and costs have the same importance for customers. Some are more important than others. That is why we ranked them in declining order of importance in the Product Value Profile. In general, customers have a few very important value drivers that play a critical role in their decision process. Others are less important. How many is "a few"? In the previous section we suggested one to four. But is it two, three, four or even five? To answer this question, we need to be able to assign a weight to each value driver. A powerful quantitative method does exist to find these weights: it is called conjoint or trade-off analysis

Conjoint analysis is a powerful market research tool that produces numbers that represent the value customers place on products and product features. It is powerful because it places the customer in a position to choose between products and, therefore, make trade-offs. In one implementation approach of the tool it asks the customer to rank products — each represented by a list of key attributes or value drivers — in order of preference. Based on these rankings, and using statistical analysis tools widely available in commercial packages, the researcher can find how much the customers value each attribute and each level of these attributes. The advantage of this method is that it forces the customer to make trade-offs between value drivers, and consequently it provides the most accurate representation of the customer decision process, assuming the customer will behave in real life as he or she does in the simulation. Conjoint analysis requires the participation of customers, so it is not cheap. If you cannot afford to do a full conjoint analysis, a more approximated approach is for the product team to allocate 100 points among the CVDs based on the knowledge it has of the customers. Beware! This approach may be subject to large errors and biases, as most product teams have a tendency to overrate their products and underrate competitor products. They also have a bias for the attributes they consider important. Nevertheless, it retains one important advantage: the need to make trade-offs. If the team does not have the required knowledge, another approach is to take as weights the inverse of the rankings as done in the previous section (CVD#1 would receive a weight of 10, CVD#2 a weight of 9, and so on), or even to assign a weight of 10 to the top three, a weight of 5-7 to the next four, and a weight of 3-5 to the last tier.

> Important
> The authors have developed a simple conjoint analysis tool that allows a marketer who is new to the technique to get familiarized with it. It is free for the asking.
>
> Send a request to conjointanalysis@rapidinnovation.com indicating your name, company and application.
>
> For an introduction to Conjoint Analysis, see www.sawtoothsoftware.com/conjoint-analysis-software.
>
> For packages that you can buy to perform Conjoint Analysis, see www.sawtoothsoftware.com and www.spss.com/conjoint.

Level 3 Detailed Process: Calculate the Product Value Index

Where You're At

Calculating the Product Value Index

Once the importance of each value driver has been determined, you can calculate the PVI of each product by calculating the weighted sum of the score for each value driver with the accompanying template.

See Also: Product Value Index Example

1. **List the Most Important CVDs**

Start by listing the most important Customer Value Drivers in declining order of importance in column C. Column A is pre-filled with ranking numbers 1-10. This is the same order as the one you did for the Product Value Profile. Double-check that you have more than purely technical benefits among the 10 value drivers.

2. **Rank the Importance of Each CVD**

For each Customer Value Driver, enter the importance of this value driver in column B so that the total of all "Importance" is equal to 100. Also enter the category of each value driver in column D.

3. **Enter Your Customer's Ratings**

For Competitor A, which in most cases will be your company's product, enter how your target customers would rate your product on a scale of 1-10 in column E. This number can also be the result of a conjoint analysis or a similar approach.

4. **Determine the Score**

For each line, multiply the Rating in column E by the Importance in column B and enter a Score in column F.

5. **Determine the Product Value Index**

Add all the values in column F and enter the result at the bottom.

6. **Compare Competitors' Offerings**

Repeat the process for the other products you want to compare.

Faced with a Need

Example: Product Value Index

Template

(A) No.	(B) Rank Importance	(C) Customer Value Drivers in Order of Importance	(D) Category	Your Product (E) Rating (1-10)	Your Product (F) Score	Your Competitor's Product Rating (1-10)	Your Competitor's Product Score	Differentiation
1	29%	Ability to address existing and anticipated technical problems.	Ben/Tech	8	2.32	10	2.90	**BEST** You need to be the best in satisfying the most important Customer Value Drivers (CVDs).
2	18%	Positive experience with the product.	Ben/Emo	9.5	1.71	8	1.44	
3	15%	Positive experience with the company.	Ben/Emo	5	0.75	9.5	1.43	
4	12%	Ease of operating and maintaining the system through its life.	Ben/Tech	7	0.84	8.5	1.02	**COMPARABLE** You can be comparable to competitors in the second tier of CVDs.
5	8%	Increased asset utilization.	Ben/Eco	8	0.64	4	0.32	
6	5%	Ease of integration into company processes and operations.	Ben/Oper	9.5	0.48	4	0.20	
7	5%	Cost of ownership (CoO).	Cost/Money	8.5	0.43	6	0.30	
8	3%	Impact on the organization if something goes wrong.	Cost/Risk	9	0.27	7	0.21	**INFERIOR** You can be inferior in satisfying CVDs that are the least important, provided your performance is still acceptable.
9	3%	Increased revenue/throughput.	Ben/Eco	7	0.21	4	0.12	
10	2%	Purchase price.	Cost/Money	6	0.12	4	0.08	
All	**100%**	**Product Value Index**			**7.77** (G)		**8.02**	

The PVI (Product Value Index) is a numerical expression of value. By quantifying the value that a product represents to a target segment, you can determine how the offering of one company ranks compared to that of another. The PVI is calculated by first multiplying the Importance percentage (B) by the Rating (E) — this generates the Score (F) for each CVD — and then totalling all the Scores to create the PVI (G).

In this example, Your Product has a PVI of 7.77. This score is relevant only in comparison with another PVI. The example shows that Your Competitor's Product has a PVI of 8.02. Though not a big difference, only one product can win in each situation. In addition, Your Competitor's Product has better ratings for the top value drivers, increasing its chance to win in side-by-side comparisons.

The Whole Product: The Product Versus the Solution

From the Customer's Perspective

We have used the words "offering," "product" and "solution" to designate how a company delivers value to the customer. These words are widely used, and you all think of these words as something your company offers. But remember that you have to think like a customer. How do customers think about your offering, your product or your solution? Have you ever asked your customers why they buy from you? If you have, you probably have noticed that they do not give the same answer you would. Typically, they mix product features, benefits, solutions and also intangibles. They might even mention that they like the people with whom they interact. Marketers typically don't consider this emotional aspect, or, if they do, it is something they cannot do much about. For most product planners the only acceptable answer is a better product. We call the customers' perception of what they buy or consider buying the "whole product."

Definition — The Whole Product

The necessary combination of tangibles and intangibles that, when seen through the customer's value driver filter, satisfies his or her Compelling Reason to Act (CRTA). The perception of what customers buy or are considering buying.

The whole product is the combination of tangibles and intangibles that, when seen through the customer's value driver filter, satisfies his or her Compelling Reason to Act. The whole product is what customers actually buy.

Who in your company is in charge of the whole product? Nobody? Several people? If you want to deliver on what the customers value, it may be time to think about having a "Whole Product Manager."

Ideas for Action: Define Your Whole Product

While the concept of the Whole Product — or Total Solution — is simple, the implementation is difficult, particularly if your company is technically-driven and the banner specs take priority. Start by using the Dimensions of Value shown at the start of this guidebook. Use the dimensions to establish a conversation about solutions and not just products.

Start using Use Cases to describe what your customers are trying to achieve. This step will vividly remind you that your customers are trying to solve problems, not just use a product. With some effort, you may be able to video tape or photograph the various steps used by your customers to perform their duty. This will also create a different conversation, away from product and towards solutions.

See Also — Dimensions of Value

Customer Experience

Good Products are Not Enough

Additional Areas to Add Competitive Advantage

We have seen earlier that there are 10 areas where you can create value. Most companies focus on the technical advantage of their products. But today the technical area is less and less a sustainable competitive advantage, as most product improvements can be rapidly reverse-engineered. So companies and scholars have been looking for additional areas that can create an advantage. One of those areas is the emotional benefits. Emotional benefits are easy to understand when we talk about consumer products, but for business and professional products it is less obvious. It has been said for a long time: "People make emotional decisions that they then justify rationally." But up until now, this domain was left unexplored. Now we must begin to understand a little better how emotions can add value to business customers.

Faced with a Need

Creating Products and Companies People Love

"Customer Satisfaction is Worthless; Customer Loyalty is Priceless"

Behind the provocative title of this book by Jeffrey Gitomer stands two recent fundamental marketing findings: satisfied customers are not necessarily loyal to companies, and satisfied customers actually change brands easily. Why? Because most people will say they are satisfied when actually they have nothing great, good or bad to say about your company or your product. Satisfied now means neutral. So measuring customer satisfaction is an unreliable indicator of future success.

Only Three Questions are Really Important

1. Will the customer buy from you again?
2. Will the customer recommend you and your product to others if asked?
3. Will the customer become a champion for your product?

Put another way, **LOYALTY** and **ADVOCACY** are the most valuable traits to nurture in customers. Loyal customers and advocates keep purchasing from you and say good things about you to others. How do you generate loyalty and advocacy? The short answer is that you need to WOW the customer, because as human beings we are loyal to those who can make us experience unusually pleasurable feelings when interacting with them. Furthermore, people want to repeat pleasurable feelings! Apple customers have long been known for their nearly unconditional loyalty to their brand. So are Coca-Cola customers. But they are even more than loyal; they have become advocates for these brands, soldiers who spread the good word. We all have brands and companies to which we are loyal because we have pleasure interacting with them. Generating a strong emotional connection with a brand is the key to generating loyal customers and advocates. That's why more and more companies in the industrial and business sectors are focusing their efforts on the creation of pleasurable experiences.

Think a moment about a good or a bad experience you had, and describe it. For example, you went with friends to a restaurant, or you went hiking with your children, or rented a car. When you describe this experience, you will notice that you describe a succession of events. When you express the reason why it was a good or a bad experience, you describe feelings, emotions and other intangibles you value. And you long to repeat those good experiences.

Creating Pleasurable Experiences

Can a company create pleasurable experiences for its customers? Absolutely! There are two areas in which companies can improve: experience with the product and experience with the company.

Experience with the Product

Here again, think of Apple. The first Macintosh created a revolution in human interface that, for many people, was so pleasurable that they became evangelists and soldiers serving the cause of Apple against IBM and Microsoft. One key way to create pleasurable product experience is through the creation of a human interface that is easy, intuitive and pleasant. The second is through the aesthetics of the product and design. A beautiful object is something that makes you feel good. Not surprising that design is growing in importance in many companies. To think "experience" rather than "product" is not the easiest thing to do.

One way to force you to think experience is to add the suffix "ing" at the end of your product name. A calculator would create a calculating experience, an oscilloscope would create an oscilloscoping experience, a bicycle would create a bicycling experience. It does not work with all product names, but just thinking this way causes you to consider the sequence of events that make up the experience, rather than thinking about products and product features.

Faced with a Need

Experience with the Company

It is rare that customers have a single interaction with a company. Most likely, they will have pre-sales and post-sales interactions. For example, they will look for information on your Web site, interact with salespeople, have a technician come install the unit and have invoicing and service-related questions. Each of these experiences can be pleasurable, neutral or frustrating. Think of being put on hold for 30 minutes before reaching a customer service representative. These interactions are called "moments of truth" because it is at these moments the customers build their opinions about your company. Your company's Web site, employees, processes and communication messages are all contributing to these moments of truth that can create loyalty, neutrality or rejection. Creating pleasurable experiences should therefore receive the utmost attention within a company. It's a responsibility that seldom rests on one single person, but probably should. Maybe we will see one day a Customer Experience Manager who coordinates all the activities that create moments of truth.

Change Your Focus from Product to Value

To succeed in today's competitive world, marketers and managers need to change their mind from a product-centered focus to a value-and-experience-centered focus. This is the same change we have been advocating when we discussed value — not surprising, since a good experience creates emotional value. Of particular importance is the need to think of value rather than features and to consider the customer as an active participant with feelings and emotions.

Important! To increase customer value, change your mindset from improving features to improving the customer experience.

Focus	Product-Centered	Experience-and-Value-Centered
Customer role	Passive	Active
Customer involvement	Remote	Central
Communication	Product-centered	People-centered
Marketer's focus	Product features	Value, experience with product and company, and moments of truth
Sales	Push products	Focus on customer's buying experience
Source of value	Technical value	All dimensions of value, in particular emotional value
Product design and aesthetic	After the fact	Integrated right from the beginning
Indicator of product success	Fit to function	Loyalty and positive recommendations
Indicator of company success	Sales and market share	Share of mind and loyalty

"Product-Centered" is conventional thinking; whereas, "Experience-and-Value-Centered" is a new and improved approach. Your organization needs to make a commitment and develop a roadmap to make the transition from Product-Centered to Experience-and-Value-Centered, starting with the highest leverage items on the list.

The Master Process

NEW TOPIC

Introduction

The journey from soliciting customer input to actual products is indeed a long one, which at times may feel overwhelming, especially if it's the first time. Fear not — do not hesitate to embark upon the process. The first time is not likely to be perfect, but it will yield visible results. As you run through the process, things will become clear and even simple. You've reached the end of the road when the entire process of obtaining and using customer input is no longer a novelty for you and your team, but rather a well-ingrained habit and simply the way you do things.

Definition

The Master Process

A visual tool to help you navigate through the major milestones — a "bird's-eye view" mapping of the entire system.

The Master Process

The Master Process is a mapping of the entire system. It gives you the proverbial bird's-eye view of the journey from start to end. The Level 1 Master Process is a visual tool to help you navigate through the major milestones. The details for each milestone are presented in the following chapters in the forms of Level 2 Main Processes and Level 3 Detailed Processes.

- You can use the Master Process to understand all the major elements that ensure the voice of the customer is integrated into your new products.
- It is helpful to communicate with the rest of your team how you obtain, process and incorporate the voice of the customer.
- It is a tool to communicate with management and the executives of your organization. You can vividly show them the entire journey, and talk about budgets, time and resources.
- It serves as the "you are here" map that can help you get oriented and regain your bearings, particularly when the entire journey may take months.
- It can be helpful in ensuring that the engineering or design organization incorporates the voice of the customer milestones into the engineering schedule.
- Use the Master Process presented in this guidebook as a template to create your own Master Process, one that includes the language used in your organization and your current product development process.

Make the Customer Part of Your Team

The Master Process vis-à-vis the Product Development Plan

The Product Development Plan (PDP), also called the New Product Development plan (NPD), is the map that takes the product from idea to the customers. It drives all activities in the development of a new product, including the voice of the customer. It is, therefore, critical to connect the Master Process with the overall plan that your organization uses.

Early in the planning phase of new products, a PDP, NPD or MRD (Market Requirements Document) is generated, likely by your engineering organization in collaboration with marketing. It is at this stage that the Master Process for the voice of the customer must be developed and incorporated into the overall plan.

Often, the activities for customer input are not incorporated early enough in the overall plan, which results in zero time allowed for customer interaction. To avoid this contingency, ensure that from the start proper time and resources are accounted for, documented and approved.

The Schedules

In new product development there are always schedules, which are developed to ensure timely delivery of the new product; for example, the engineering, marketing and manufacturing schedules. Ensure that these

Essential

"If customer interaction is not in the engineering schedule it will not happen!"

schedules include the right amount of time for customer interaction; often this is left out. Pay particular attention to the engineering schedules, and ensure that there is a line item with voice of the customer, or customer interaction or customer input included. If it is not in the engineering schedule it will not happen!

> Essential: "Obtaining and using the voice of the customer takes time. You will need to change the schedules to accommodate the additional time needed to get the job done."

Obtaining and using the voice of the customer takes time, but that time will be more than compensated for, when later in the development you will be able to eliminate excessive reworks, avoid unproductive planning sessions and streamline procedures. And if you are not currently doing it, or are only doing parts of it, you will need to change the NPD plan and the engineering schedules to accommodate the additional time needed to get the job done.

The Master Process for the True Voice of the Customer

Customer intimacy is the ability to see the world through the eyes of the customer. Customers have a different perspective of the world, your product and their needs. Thus, it is critical to not only understand their needs but actually appreciate the world through their eyes.

While we all profess to be customer-driven, few of us can articulate the process for obtaining customer input and translating it into effective action for our projects.

The system to achieve the potential of customer intimacy in a new product development effort is made up of four different phases:

1. **INVESTIGATE**.
2. **INTERPRET**.
3. **INNOVATE**.
4. **INCORPORATE**.

The 4-I Framework

An Overview of the Four Phases

The tables on the next few pages show the Master Process as it separates into its four phases. These tables gives you a description of each phase, the major steps for each phase, and the methods and tools used in each phase. In addition, you will discover how Voice of the Customer (VOC) methods map to the Master Process

The Output

There is a clear set of deliverables or outcomes for each of the four phases. Clearly defined outputs will help your team focus on the task at hand. Each phase builds on the previous one to form the Master Process. Each phase also has a specific document, which is the manifestation of the output of that particular phase.

Many organizations already have documents that represent the output; for example, the MRD (Marketing Requirements Document), or the ES (Engineering Specifications). You should make an effort to reconcile the names to match the four phases. Pay particular attention to documents that do not yet exist in your organization, and you may also need to modify existing ones.

The Master Process

VOC and the Product Development Cycle

Most organizations have a process to develop new products. Often called the NPD, it is the process intended to guide the development team from start to finish — the point where the product is ready for purchase. In some cases there may be no documented process; regardless, there is a calendar-based, step-by-step progression. In addition, there are many product development methods, such as Stage Gate, Agile, Waterfall, Iterative, etc. They all require close interaction with customers.

See Also
Life Cycle Models Appendix
Stage Gate, Agile, Waterfall, Iterative, etc.

While there are significant differences of product development processes across industries, the basic premise is common to all: the ability to develop a product on-time and on-profit. For example, software and hardware development processes are different, but the need for customer interaction is critical in both cases.

Make note of when in the process you will conduct customer interviews, other forms of customer interaction and market research. Also note that doing the wrong type of interaction may be detrimental to the development of your new product.

NPD (New Product Development) Process

This diagram shows a typical product development process — yours will not look exactly like it, but the major milestones should be very similar.

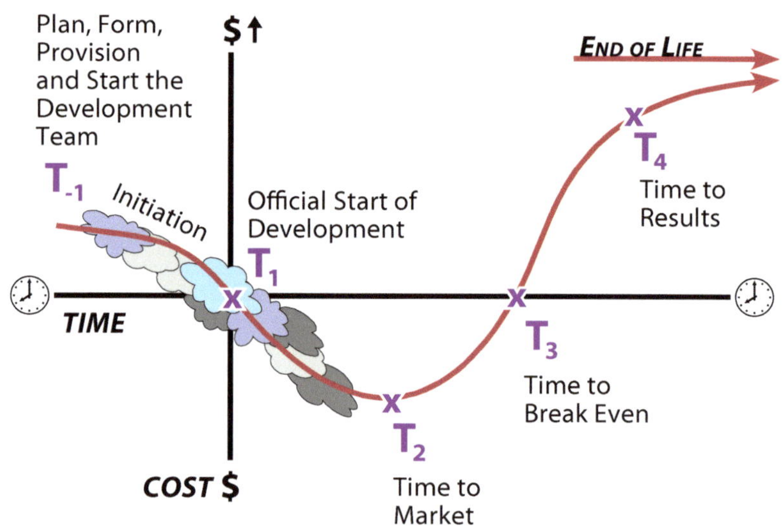

The x-axis represents time, while the y-axis represents costs.

The Initiation phase (T_{-1}) allows you to organize the development team, start the documentation, obtain initial approvals, etc.

The Official Start of Development phase (T_1) is generally marked by an extended meeting plus special documentation that enables the expenditure of resources towards the eventual introduction of the new product. While some funds would have been spent during the Initiation phase, the Official Start of Development marks the expenditure of substantial resources.

Time to Market (T_2) is the official date for the announcement of the new product and the completion of the Development phase. At this point, the new product is available for pre-purchase and early-delivery.

Time to Break Even (T_3) is when you recapture your initial investment and are on the verge of making a profit.

Time To Results (T_4) is a point that you select to validate the business success of the new product. The point is selected at the start of the Development phase or, better yet, during the Initiation phase, It's a financial metric that validates the business model in your Product Proposal. Companies establish the Time to Results because they can't afford to wait several years to measure the success or failure of a new product.

Interaction with your customers starts well before the Initiation phase. You must understand the needs of your customer in order to prepare your Business Case — the proof that your new product will succeed in the market place.

Your organization must have a framework for customer interaction — a formal and disciplined way to obtain customer feedback and input all phases of development better and faster than your competitors.

The Master Process

Applying VOC During Various Stages of NPD

Note that other forms of customer interaction and market research (e.g., usability testing, focus groups and surveys) are occurring at the same time. Note that each phase has a set of outcomes and clear documentation. Also note that Engineering R&D and Marketing must collaborate in all phases to ensure clarity of understanding and accelerate Time-To-Market.

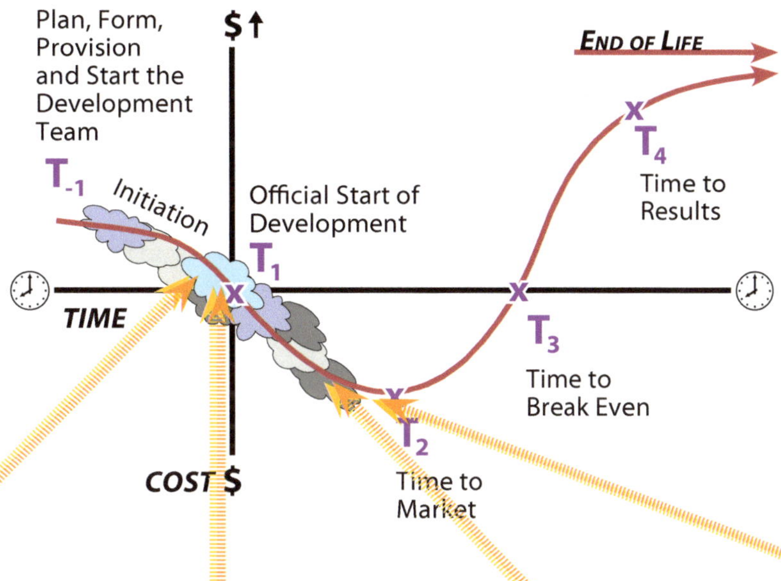

Planning Phase

The period of time spent understanding major market and technology trends, also called Concept Development Phase.

Preferred Approach

A series of in-depth interviews with customers who understand the new trends, combined with primary market research.

Documentation

Business Requirements Document or the Business Case Document. Marketing delivers market segmentation, product categories and the financial case.

Product Definition

The start of the development process — at the end of this phase the definition of the product is completed, but not necessarily finalized.

Preferred Approach

A second series of customer interviews, this time in full collaboration with engineering. Marketing also conducts secondary market research.

Documentation

Market Requirements Document (MRD). Marketing delivers a full set of customer requirements after a series of customer interviews and other market research.

Product Design

The new product is being designed, the definition is finalized.

Preferred Approach

Customer interviews that focus on a particular aspect of the product; for example, user interface (GUI).

Validate that the customer requirements have not changed and the message from the customer has not been altered during definition.

Documentation

Marketing delivers specific input from customers on specific set of questions.

Final Testing and Production

At this point, the product is in final refinement and being prepared for public announcement.

In some industries, the beta versions are tested with customers for refinement — not for major changes.

Preferred Approach

Specific testing with the customers, beta programs and early demonstrations.

Documentation

Marketing delivers messaging, positioning and sales training.

Marketing delivers a prioritized list of minor changes as a result of the beta program.

The Master Process

Applying VOC to Optimize Customer Interaction during NPD

Interaction with your customers is a never-ending effort — it does not stop when you finish a product development project.

Furthermore, there are several types of customer interaction, and many occur simultaneously. For example, your marketing organization may be defining a new market (Market Sensing), while at the same time developing a new product for an existing market (ideation plus development). There must be clarity in your organization as to which type of customer interaction is needed. There is a clear need to develop a solid framework for customer interaction. We suggest that you develop a Key Competency (Core Competency) to understand your customers better and faster than your competitors.

- **MARKET SENSING**: This phase focusses on the major trends in the market place and the emerging market needs. VOC focusses on the industry leaders, the industry opinion makers and technology development.
- **STRATEGY MARKETING**: VOC is used to determine the direction and requirements of an entire market or market segment. VOC helps identify product categories and new products that will be needed by the market in the future.
- **IDEATION PHASE** : VOC at this point is product specific, but open to the full scope of customer requirements.
- **PRODUCT DEVELOPMENT** : Some of the VOC is validating that the customer requirements have not changed and the message from the customer has not been altered during definition.
- **PRODUCT INTRODUCTION**: At this point, the product is in final refinement and being prepared for public announcement. VOC is used to continually refine the messaging, positioning and other value propositions.
- **MATURITY AND PHASE OUT**: VOC at the end-of-life focusses on exit costs, barriers to switch to new product categories and identifying the penalty customers will pay if the product is no longer available.

Process Map: VOC and the Full Product Life Cycle (PLC)

Product Development System

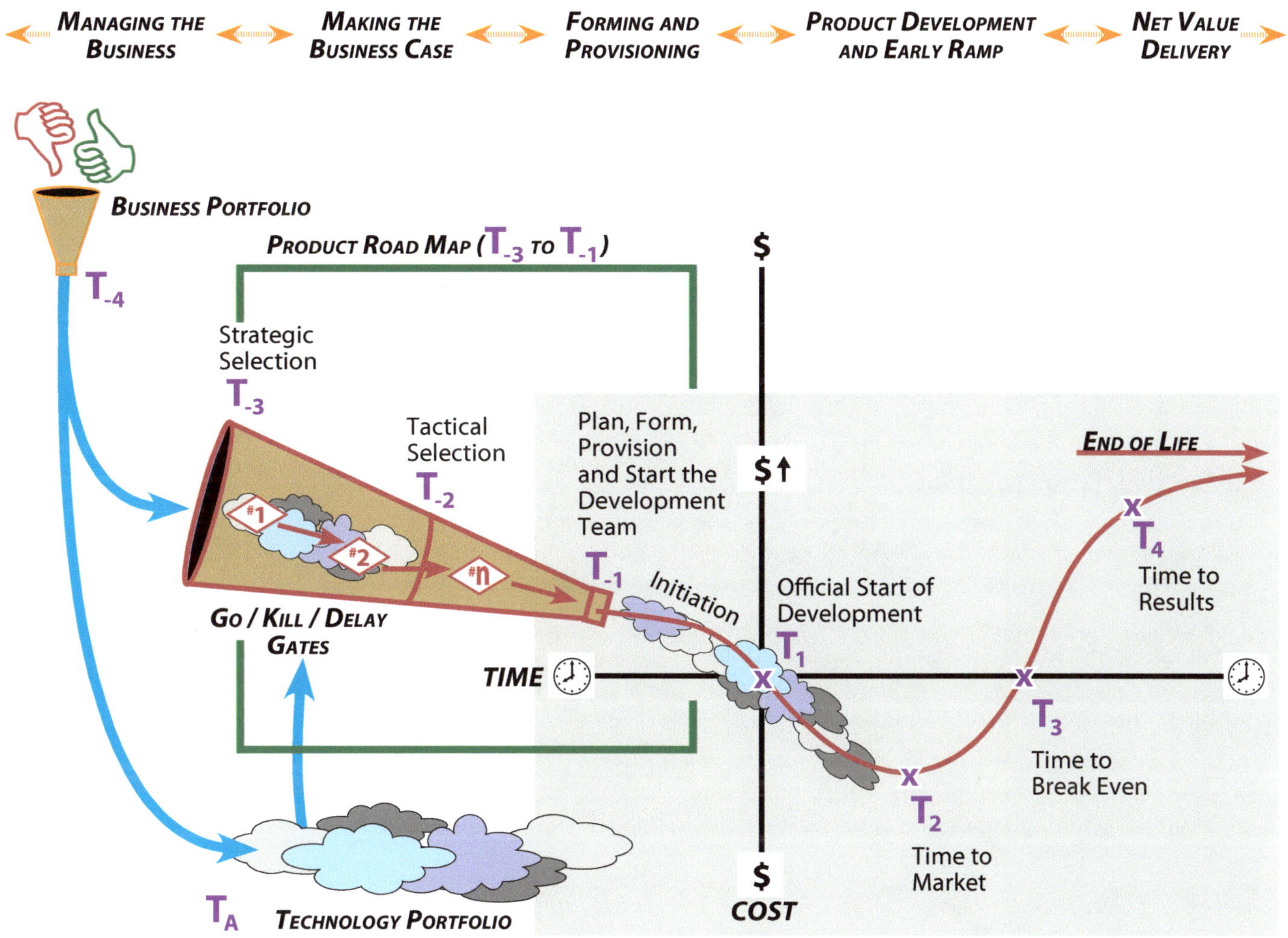

The New Product Development (NPD) process is part of the Full Product Life Cycle (PLC). It's important that you and your team understand the dynamics of running your business and how decisions are made — your development team is part of a system where priorities are set based on the business merits of projects.

The Business Portfolio (T_{-4}) (also called the Corporate Portfolio) is where the high-level decisions are made by the executives of the company; for example, Key Strategies to manage growth, acquisitions, mergers, large investments, etc. When the Business Portfolio drives the entire system, one can say there is Strategic Alignment in the corporation.

The Product Roadmap (T_{-3} to T_{-1}) (also called the Product Funnel or Product Portfolio) represents the decision making process to prioritize the NPD to meet the business objectives of the organization in support of the Business Portfolio. The Product Roadmap has two distinct phases: the strategic selection — decisions to align with the business portfolio — and the tactical selection — assessing the development resources to ensure a successful rollout of the new product. If your company has several divisions or business units, then each will have its own Product Roadmap.

Generally speaking, the Technology Portfolio (T_A) is under the purview of the Chief Technology Officer (CTO) or Central Engineering for the purpose of developing and staging the new technologies needed to support the Product Roadmap.

The Master Process

More on the Product Development Process

Regardless of the amount of detail in your New Product Development process (NPD), the key message is the absolute need to involve your customer throughout the process. Your timeline (schedule) must vividly include the actual interaction that you and your team will have with your customers, keeping in mind each of the phases in your product development cycle.

The same applies to whatever methodology you select, from Agile to Waterfall to Scrum, the need for proactive interaction with your customer does not go away. If your methodology does not require proactive interaction with your customers, change the methodology or adopt a different one.

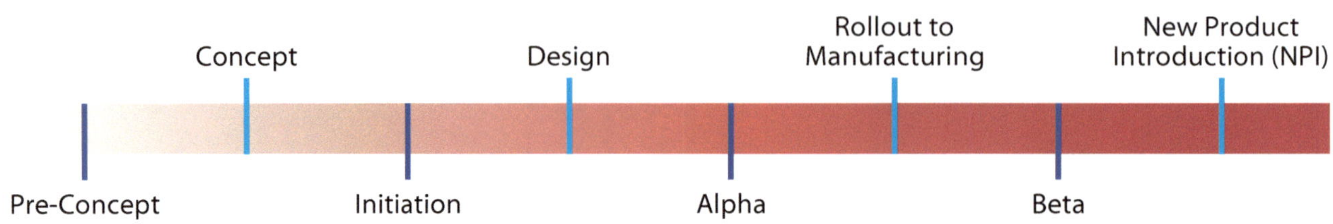

Software Development

Software development differs from hardware in several ways, but the need to deliver value to the customer does not change. The authors recognize that the processes do have differences, but at the core they are very similar. Their implementation may differ, but at the strategic level they are very similar.

If anything, newer software development practices require frequent and more extensive interaction with customers. In the end, the need to clearly understand customer requirements beyond technical needs is as important for software as it is for hardware. Both provide a solution to the customer's problem, and usually the solution is not just technical, regardless of software and hardware.

Tools used to enhance the understanding of customer needs in software include use cases, Business Rules methods and process mapping. The truth is that very similar tools are used in the hardware area. And, they are all intended to better understand the customer's problem and define the most value.

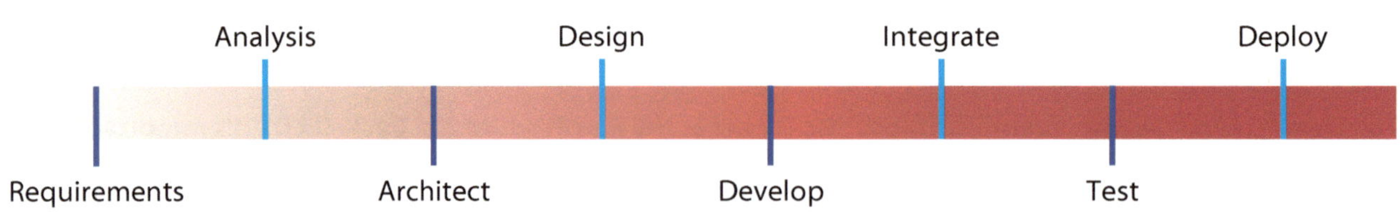

Level 1 Master Process: The 4-I Framework

Where You're At

— Voice of Your Customer —

The Market Requirements Document (MRD) Process

1 Investigate

Investigate ➲ Voices

Enabling customers to unveil their needs, goals, issues and challenges, to be expressed as customer voices.

2 Interpret

Interpret ➲ Requirements

Converting customer voices and input into a prioritized list of requirements.

Important! Before you can create value, you must first discover what your customers' requirements are for value.

Where You're At

— Value Creation —

The Product Requirements Document (PRD) Process

3 Innovate

Innovate ➲ Product Concepts

Translating customer requirements to innovative product concepts.

4 Incorporate

Incorporate ➲ Solutions

Developing solutions based on product concepts that faithfully respond to customer requirements.

The Master Process

Level 1 Master Process: The 4-I Framework

It's important that you internalize the definitions of the 4-I Framework because that will help you lead your development program. Also, ensure that your team understands the implications of this master process.

You may have different names for your existing documentation or even documentation that isn't addressed in this guidebook. If so, compare and keep the best combination to meet your objectives. The key point is that each phase must have a tangible and verifiable document to demonstrate completion.

— Voice of Your Customer —

1 Investigate

The disciplined approach to gathering and documenting unfiltered customers, articulated and unarticulated needs through a variety of techniques that always include in-depth interviews and some level of observation of the customer's world.

VOICES

Output

At the end of this phase you should have captured (documented) the voice of the customer: the raw data from each of the interviews that you and your team conducted.

Documentation

This should be in the form of notes, photos, video and audiotape, plus the vivid memories that you and your interview teams captured.

2 Interpret

The process of organizing the customers' data and processing them into a set of prioritized and documented requirements that clearly express the customers' unsatisfied needs.

REQUIREMENTS

Output

At the end of this phase, you should have a clear, prioritized and approved list of Customer Requirements.

Documentation

Generally, this list is relatively short, between two and 10. In some cases it may be longer, but we encourage prioritization to reduce the number of the requirements, which capture the value expressed by your customers.

— Value Creation —

3 Innovate

The process of transforming requirements into one or a few product concepts that will address the customers' articulated and unarticulated needs.

PRODUCT CONCEPTS

Output

At the end of this phase you should have innovative solutions that address the requirements expressed by your customers.

This phase is where engineering and marketing collaborate to find the solutions and to innovate.

Documentation

The documentation should be a prioritized list of solutions, or a collection of product features that in aggregate (cumulatively) provide a solution.

4 Incorporate

The disciplined refinement of the product concepts into a single one validated by customer feedback, and its full articulation into a set of features and specifications that engineering can implement.

SOLUTIONS

Output

Naturally, the desired output of this phase is a profitable product.

Documentation

On the more practical side, the output of this phase generally is an engineering document, which takes into account the technical trade-offs, the timeline, design costs and other parameters. Engineering and marketing collaborate to ensure that the customer always wins; that is, that the product always has maximum value.

The Market Requirements Document (MRD) Process

The Product Requirements Document (PRD) Process

Level 1 Master Process: The 4-I Framework

High Level Process

The steps contained in each of the four phases have been diagrammed in further detail to demonstrate the progress from the voice of the customer all the way to engineering specifications.

— VOICE OF YOUR CUSTOMER —

1 Investigate

STEP 1. PLAN YOUR VISIT

- Develop the Charter
- Develop the objectives and purpose statement. (Why are we doing this?)

STEP 2. ORGANIZE YOUR VISITS

- Prepare the sample selection. (Who are we going to ask?)
- Define and test the discussion guide. (What and how are we going to ask?)
- Identify the team for the interviews. (Who will do the asking?)

STEP 3. INTERVIEW YOUR CUSTOMERS

- Conduct the interview. (The critical moment)
- Capture the information faithfully.
- Debrief #1 Debrief immediately after every interview. (What did we learn from this specific interview?)
- Prepare for the next interview. (What do we need to improve for the next interview?)

2 Interpret

STEP 4. PROCESS THE INFORMATION

- Consolidate your notes from the interviews.
- Highlight your notes.
- Create Yellow Stickies. (From your notes, transfer relevant information to Yellow Stickies.)

STEP 5. CONDUCT THE POST-VISIT MEETING

- Schedule the extended meeting. (Well in advance, to ensure complete attendance — everyone who went on the interviews.)
- Do the Affinity Diagram. (Sort information by major categories.)

STEP 6. CREATE IMAGES AND STATEMENTS

- From the Affinity Diagram, create Images and Statement, and the opportunity for your team to dialog.

STEP 7. DEFINE AND DOCUMENT REQUIREMENTS

- Establish good requirements. (The product shall have the capability of... The user shall have the capability of...)

— VALUE CREATION —

3 Innovate

STEP 8. MARKETING AND ENGINEERING COLLABORATE

- Develop innovative solutions to your customers' problems.

4 Incorporate

STEP 9. DEFINE SPECIFICATIONS

- Prioritize the innovations from the previous phase.
- Prioritize the attributes of your new product.
- Prioritize the features of your new product.

STEP 10. DEFINE ENGINEERING SPECIFICATIONS

- Define engineering specifications. (Engineering does the heavy lifting in this phase, in collaboration with marketing, to ensure the preservation of the voice of the customer.)
- Incorporate into the design documentation. (Engineering specifications.)
- Monitor progress, and adjust to reality.

The Master Process

Level 1 Master Process: The 4-I Framework

Definitions

Definition

Attributes

Not the new features, but the overall qualities that will characterize your product (e.g., compact, or easy to use). It is the aggregate or cumulative result of all the features of your product, those that cause your customers to express the overall character of your new product.

Definition

Heuristics

Methods of solving a problem for which no formula exists, based on informal methods or experience.

Definition

Images

The verbatim description of the world as your customer sees it. The expression of how it feels to confront the challenges and opportunities facing your customers. See the Post-Visit Process (PVP).

Definition

Imaging

A vivid mental picture that describes, in emotional terms, the problem, challenge or issue confronting your customers. Preferably, it is a verbatim phrase, sentence or narrative from your customers.

Definition

Innovate

The ability to create an environment that enables new ideas, creativity and breakthrough thinking. Typically characterized by extended meetings where an open dialog and the world of the customer come first. The ability to bring new value to the customer. It is not innovation until your customer says it is!

Definition

Neurolinguistics

The linkage between language and the brain. It is the brand of linguistics that explores how the brain encodes language.

Definition

QFD (Quality Function Deployment)

A framework to manage customer input in a way that drives the development of a new product; a tool that puts the input from the customer first. It incorporates the Voice of the Customer into the overall product development plan.

Definition

Semantics

The study of the meaning in language. It is the study of how meaning in language is created by the use and interrelationships of word, phrases and sentence — the way words are interpreted regardless of their dictionary definitions.

Important

At the earliest stages of planning for the new product, ensure that you have the resources allocated for interaction with your customers; i.e., your customer visits. Also, ensure that all schedules, including the engineering time-line, clearly show the time allocated for customer interaction.

See Also

Post-Visit Process

For comments, questions or additional copies please e-mail the author: voc@rapidinnovation.com

Level 1 Master Process: The 4-I Framework

Methods and Tools Used

Some of the methods and tools commonly used in each of the four phases of the Master Process.

— Voice of Your Customer —

1 Investigate

- ❑ Sampling methods
- ❑ Heuristics
- ❑ Ethnography
- ❑ Naturalistic observation
- ❑ User profile
- ❑ Activity mapping (Process mapping)
- ❑ Environmental analysis
- ❑ Use analysis (Modal analysis)
- ❑ Scenario mapping
- ❑ Focus group techniques
- ❑ Time and motion studies
- ❑ Usability testing
- ❑ Use cases
- ❑ Anthropological methods

2 Interpret

- ❑ KJ (Kawakita Jiro method)
- ❑ Language Processing (LP)
- ❑ Semantics
- ❑ Neurolinguistics
- ❑ Personas
- ❑ Imaging

— Value Creation —

3 Innovate

- ❑ Breakthrough thinking skills
- ❑ Creativity tools
- ❑ Collaboration methods
- ❑ Advanced communication skills
- ❑ Risk-tolerance approaches

4 Incorporate

- ❑ Requirements formulation
- ❑ QFD (Quality Function Deployment)
- ❑ Documentation and document control
- ❑ Advanced communication skills
- ❑ Kano model
- ❑ Conjoint Analysis

The Master Process

Level 1 Master Process: The 4-I Framework

VOC Methods Mapped to the Master Process

Some of the most common methodologies used in each of the four phases of the Master Process.

Each of these common methods is mapped where it provides the most value; for example, QFD is useful after the requirements have been defined but not in obtaining the input and creating requirements.

The mapping is based on our research and experience.

The key point is that the four phase method is very inclusive. It encompasses all the critical functions needed to obtain, process and use the Voice of the Customer.

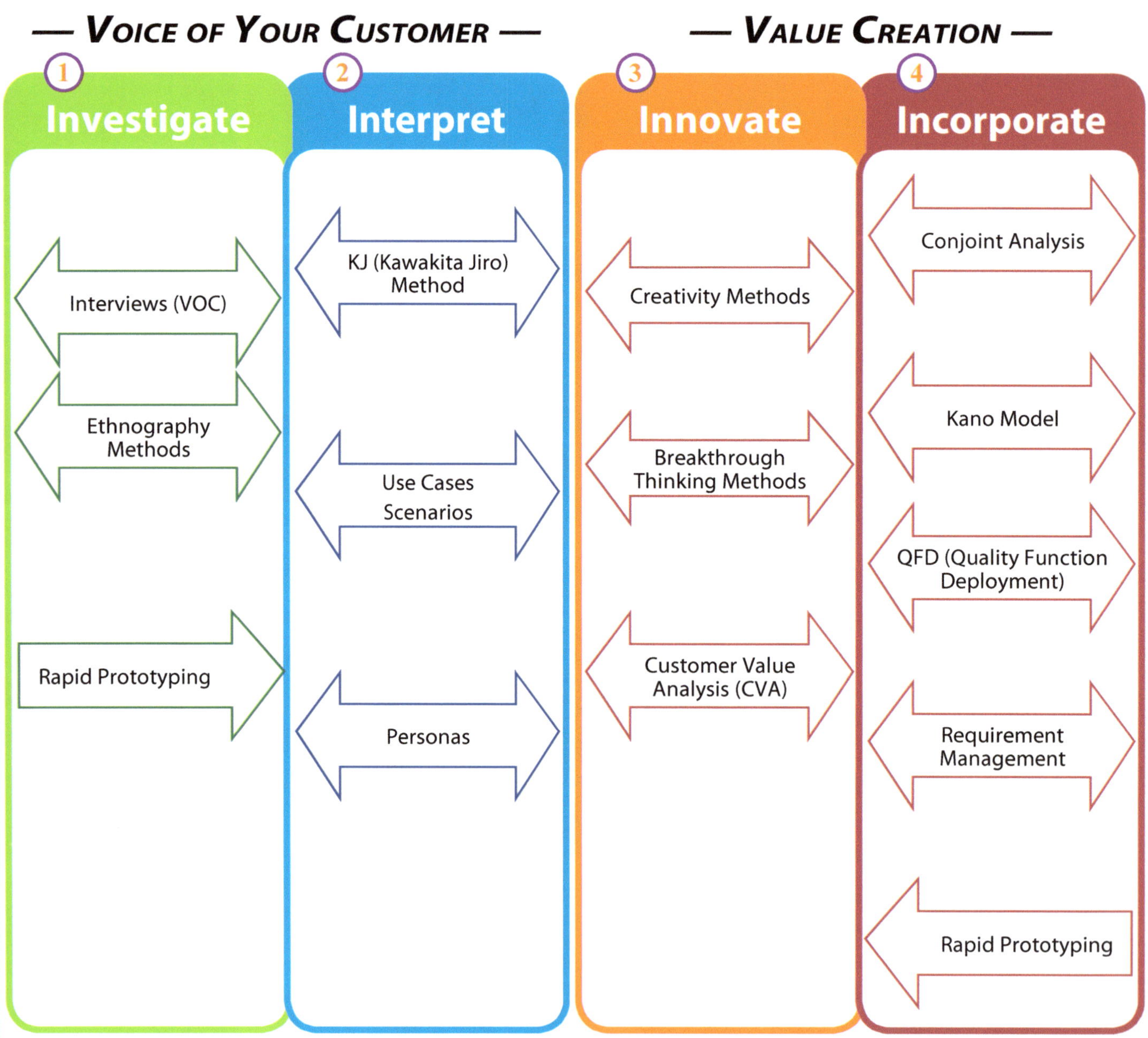

PHASE I

Investigate

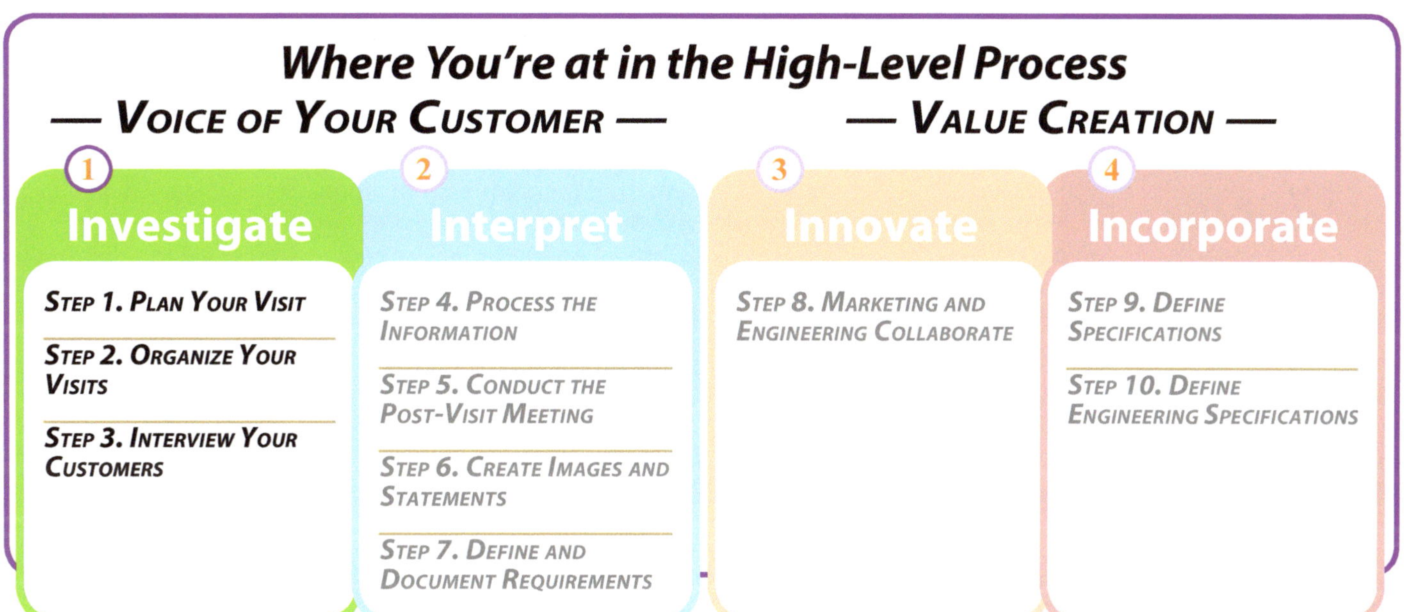

Step 1: Plan Your Visits

NEW TOPIC

The Pre-Visit Roadmap

Introduction

By definition, programmatic customer contacts are thoughtfully and meticulously planned — not casual events. A key reason that visits fail is a lack of planning. The impact of customer visits on your new product demands careful preparation to ensure maximum yield.

Remember, all you have is 12 hours — 24 at most — worth of one-hour interviews with your customers; therefore, the customers whom you do interview must be well-chosen and your time together must be well-spent in order to optimize the information you gather. From these interviews, you will derive the requirements for your next new products, so make these 12-24 hours as effective as they can be!

The Pre-Visit Roadmap tool will help you map all the steps to prepare, plan and conduct the visits. It does not cover the process of tabulating the results of all your visits — that is covered by the post-visit process.

See Also Post-Visit Process

General Rules for the Pre-Visit Roadmap

Take the time to plan — it is a worthwhile investment. The risks of poor planning include missed visits, wasting your team members' time and, most importantly, missing valuable input from the customer.

The amount of time you spend depends on the strategic importance of your set of visits and the size of the visit program. Specifically, a minor improvement to an existing product may not be strategically important, relatively speaking. Thus, your planning might be much simpler. On the other hand, if your visits include several teams going to different countries and talking to customers in various segments, the planning should be rigorous.

Include your team members in the planning — it does little good to sit by yourself and do the planning. Engaging your team will ensure a good flow of ideas, and it will create ownership and simplify communication later on. It is critical to engage those who will be visiting the customers.

It is a good idea to share your Roadmap with management and other key stakeholders after you have completed it. This will garner support.

Important Note

When we make reference to visits, that includes any type of program of customer contacts. For example, it could be a telephone program, where members of your team call selected customers and interview them. It could also be a set of meetings that you schedule during a trade show or industry conference. Thus, the term "visits" is used to include all types of programmed customer interaction, and the Roadmap will help you plan all of them.

Timing and Time

There are two critical ingredients to derive the most benefit from programmatic customer visits. The timing must be carefully planned to coincide with the engineering development schedule; i.e., you do not want to get information when the design if completed. This calls for close interaction with the engineering development organization. Watch for the major milestones, and schedule your interviews such that they have a definitive impact on your new product.

The time it takes to do a set of programmatic visits tends to be long. For example, it generally takes about eight calendar weeks from start to finish. Keep this in mind to have the customer input when it is most valuable in your development. In some cases the time may be longer, for example, interviews in other countries, that could take 16 to 20 calendar weeks or more.

But remember that the time you spend now will be recovered because the overall development process will take less time.

Essential: "A customer visit is any time you have interaction with the customer. A customer interview is when you talk to the customer as part of a programmatic set of visits."

Step 1: Plan Your Visits

Scale Your VOC Efforts to the Size of Your Project

In the sample selection, we discuss the number of visits or contacts for different types of projects. It's important to note that you will have development programs of various sizes; therefore, your sample selection will vary.

For example, for a minor improvement to an existing product you do not need to visit a myriad of customers. On the other hand, if you are developing an entirely new platform or addressing a new market, you will need to do your due diligence and talk to a sizable sample.

The consensus among the experts is that **THREE TO 24 VISITS** represent the lower and upper margins for qualitative research. The number will be determined by the strategic intent of the new product.

The Variables that Influence the Size of Your Sample

- The business objectives of your new product.
- The strategic value of the new product.
- Your budget.
- The realities of your timeline.
- The homogeneity, or lack thereof, of the market.

Important

Your new development projects will vary in size and complexity; therefore, it is important that you scale your effort to the size of your project. Avoid the cookie cutter approach; rather be flexible.

See Also **Sample Selection**

Scale Your VOC Efforts to the Size of Your Project

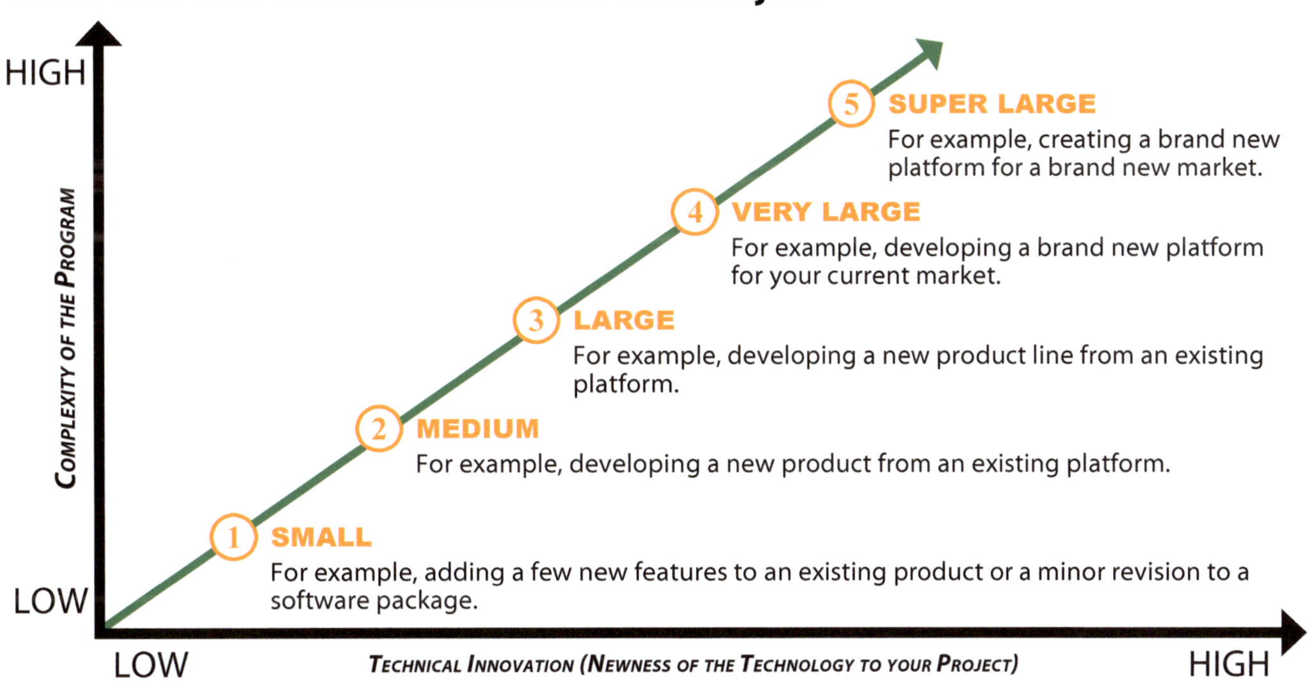

This diagram illustrates a likely relationship between your development program (Y axis) and the type of development you are performing. The key point is that not all development programs are the same — one size does not fit all. Consequently, the level of interaction with your customers must be scaled appropriately.

You can use different criteria to define the type, number and duration of customer interactions. There are several tools and templates to help you select the type and number of customer interviews in the Sample Selection section.

Step 1: Plan Your Visits

Selecting Your Team

The Framework for Ensuring that Your Next Product is Customer-Driven

First, we must establish the types of team you will need for a series of customer interviews (visits).

> **Essential** "Teams for customer interaction should always include marketing people, as well as also representatives from the engineering or technical organization."

> **Important** Remember, all you have is 12-24 hours!

1. **THE NEW PRODUCT DEVELOPMENT (NPD) TEAM** includes all the people who are working on your new product. The entire Product Development Team need not go on a set of visits because it is difficult to coordinate and most likely too expensive. These teams can be made up of as many as several hundred people in a large development program. It is the traditional way most companies organize to develop a new product.
2. **THE NEW PRODUCT CORE TEAM** is a steering committee used to drive the development of a new product from start to roll out — even if you also have a project or program manager leading the new development. This is practice from Concurrent (Simultaneous) Engineering. It has been shown that Core Team adds velocity by improving communication, improving team work and accelerating decision making. We suggest that if you do not yet have a Core Team, do establish this good practice.

   The Core Team is made up of five to seven individuals, preferably five. It is cross-functional and includes Marketing (Product Manager), Engineering (Engineering Manager), Manufacturing, Service and Repair and Finance. Other members might include Documentation, QA, Purchasing. The make up of the team varies from project to project and is determined by the needs of the development endeavor.
3. **THE VOC TEAM** advocates for the customer to ensure that the voice of the customer is obtained, processed and included in your new product. This team is also responsible for all research related to the customer, including customer interviews.

   Teams vary from three to five people and are cross-functional. Members include Marketing (Product Manager) and Engineering (Engineering Manager and some of the Engineering Leads). The Market Research Manager should also be a member, provided your company has such a function. Members meet on a regular basis to plan, review and ensure execution of all items pertaining to the Voice of the Customer and Value Creation. The VOC Team reports to the Core Team.

   This may be a new type of team to your organization. The authors believe establishing a VOC Team is a worthwhile investment. Having a small group of people always present during the entire development process, ensuring that your new product is customer-driven will improve your competitive edge and probability for success. While you might argue that this is the responsibility of Marketing (Product Manager), we have found that it takes a cross-functional team to sustain the integrity of the voice of the customer for the duration of the design cycle.
4. **THE INTERVIEW TEAMS** are made up of two (sometimes three) people who will actually perform the interviews. They may or may not be members of the VOC Team. You will have several of these teams, depending on your objectives. Membership generally comes from the Product Development Team (PDT).

   We encourage that each interview team always includes someone from Marketing and also someone from Engineering. The Engineering Managers and Engineering Team Leaders (Leads) are good candidates for the interviews. We also suggest that this is an opportunity for your best engineers to have direct contact with your customers.

Step 1: Plan Your Visits

The Framework for Ensuring that Your Next Product is Customer-Driven

New Product Development (NPD) Team

Important!

All the individuals who have direct or indirect participation in the development of the new product; for example, engineering, marketing, service, documentation, etc.

The New Product Core Team

The role of this cross-functional team is to steer the development of a new product from start to roll out. It's a practice from Concurrent (Simultaneous) Engineering.

Members of this team include marketing, engineering, service, manufacturing, finance and other roles as needed to ensure the successful development and introduction of the new product.

The Voice of the Customer (VOC) Team

A small group of individuals (three to five) who report to the Core Team and are responsible for ensuring that the new product is customer-driven. They have the responsibility of working with all the functions to maintain the integrity of the VOC throughout the entire development process.

This team is also responsible for all research related to the customer, including the customer interviews needed to identify, process and incorporate the VOC into the new product.

Members should include marketing (Product Manager), engineering, service, in-bound marketing and others as needed. They meet on a regular basis to plan, review and ensure execution of all items pertaining to Value Creation and VOC.

We have found the VOC Team to provide considerable value to your development process, and strongly believe that this team will improve the probability that your new product will be truly customer-driven.

Interview Team #1

Two (or three) persons who will do the actual customer interviews.

Interview Team #2

Two (or three) persons who will do the actual customer interviews.

Interview Team #3

Two (or three) persons who will do the actual customer interviews.

The conventional approach to integrating the voice of the customers into new products is insufficient to maintain the integrity of the VOC throughout the entire development cycle. Therefore, we propose the above diagram as a solid framework to ensure that the VOC is fully integrated into the development of your new product.

Carefully Selecting the Members of the interview Teams

Selecting the individuals who will meet with your customers as part of a set of interviews is not a trivial matter. They will be responsible for obtaining the information that can drive the success your new product. In some organizations being part of an interview team is a sign of distinction, a recognition that the individual has been identified worthy of having close interaction with your customers.

We recommend that Managers, Engineering Leads, Project Managers, and other managers and supervisors be selected. In addition, each individual must have the demonstrated maturity and experience to have close interaction with your customers. Good communication skills are invaluable. Candidates must be willing to dedicate the necessary amount of time to do a good job of obtaining the needed information.

Step 1: Plan Your Visits

Preparing for Customer Visits

Marketing and R&D see the world through different glasses.

The Stereo Vision of the Customer

We strongly recommend that teams for customer interaction always include marketing people as well as representatives from the engineering or technical organization. It is a missed opportunity if only one type performs the interaction; you are most likely to get a distorted view of the customer or miss important information.

Speaking of Visits and Interviews

In this book, we use the term "visits" to refer to customer contacts in general; any time you have a planned customer interaction it is therefore a "visit." We do this to simplify communication. The term "interview" refers to the actual meeting with one customer. Consequently, it is conceivable to have a "visit" that includes several "interviews."

The Make-Up of the Interview Team

We recommend that the teams be made up of two people, one from marketing and one from the technical community. Further, we also suggest that more than one team be formed. You want the benefits that several observers bring to the effort. For example, if you are going to interview 12 customers, form two teams and split the interviews; thus, you will have a total of four people doing 12 interviews.

If you are fortunate to be able to make more than 12 interviews — say, up to 24 — then you can create additional teams to benefit more people in your organization.

Finally, think international when forming teams to go overseas. International visits require more protocol. Work with your local marketing and sales personnel to organize your visits. Be very attentive to the local customs and the uniqueness of the cultures you will be visiting.

People from other functions beyond marketing and engineering — for example, documentation, QA, service, etc. — can be added depending on your objectives. For example, if you want input on your manuals, then it makes sense to include people from documentation, and so on.

Sourcing Customers to Interview

It is one thing to define your sample selection. It is quite another to actually find the customers to contact and then get them to agree to meet with you. This is a very difficult and challenging task for marketing professionals. Customers are very busy running their businesses. The right ones are also hard to find, which makes the task all the more difficult. We strongly suggest that obtaining customers for interviews is a special competency that every marketing organization should develop. It is difficult for a single product manager to identify, call and confirm a set of customers for interviews. This requires time and special skills. Some marketing organizations are large enough to have an individual whose responsibility is to find and confirm the right customers.

The Sales Force as a Source of Interview Opportunities

Generally, people go to the sales organization for help with customer visits. While sales can indeed be very helpful in sourcing candidates, we do not recommend having salespeople present during the interviews. Customers may not feel free to speak when "their" salesperson is in the room. There is also the risk of salespeople doing what they do well: turning the interview into a sales call. Your sales organization needs to recognize the need for direct customer interaction without their influence. On the other hand, we strongly suggest that your sales organization be kept well informed about your plans.

Another risk with using the sales force is that you may get a biased sample. Specifically, how will you get to the customers who are not serviced by your sales force? Salespeople have a territory, in which they have their good and frequent customers. Then there are the ones who belong to your competitors, those that are unhappy, etc. To balance the sample, we recommend that you use other sources for your sample beyond your sales force.

Some salespeople are very jealous of their territory and customers, and rightfully so. Their way of making a living depends on it. You should work closely with the sales manager and the individual salesperson. Present your case for customer input and the need to have quality time with the customer without the salesperson. Reassure the salesperson that you will respect the customer, and that you will brief her after your customer interview. Also, you should refrain from discussing pricing, terms and conditions for current products; let the salesperson handle these topics.

Distribution Channels

Many companies rely on alternate channels of distribution such as "reps," distributors, wholesalers, etc. In some cases the channels are reluctant to cooperate for fear of losing customers; i.e., they fear that you might make the customers a "house account." Reassure your channels that it is in their best interest to help you. The better your products meet the end user's needs, the more commissions for the channel. In some cases it will take considerable persuasion on your part. Regrettably, some distribution channels have been short-changed before, and thus they are very careful. You might want to include language related to customer visits in the contract you sign at the start of your relationship, or perhaps when your contract comes up for renewal.

Other Sources

Finally, you may have to rely on third parties, market research firms, "list-sellers," "name-brokers," etc., to get the sample you need. You sales force or alternate channels may not be able to help. They may not have the profiles you need. Remember that your loyalty is to your new product; it is in everyone's best interest to select your sample correctly the first time.

With a bit of ingenuity, you will find many sources of names. Lack of names should not be an obstacle to performing customer interviews. For example, "service tickets" provide names and contact information. People who call your technical support group may be willing to be interviewed, and you also may use your Web site to ask for volunteers (provide a simple incentive). Professional organizations are also a good place to find customers; be careful to use the right protocol so as not to abuse the rules of the organization. As you can see, you and your team can be very creative when it comes to sourcing customers for your interviews.

Important: We strongly suggest that your sales force be informed about your interaction with customers, but does not actually participate in the interviews.

As a guideline, you must have many more potential customers lined up to interview that the actual number of interviews you plan. Before selecting the final set, you must screen, which should be influenced by the type of problems, situations, etc. that you want to target. The screening task is typically performed by market researchers with a structured telephone questionnaire you have developed (in part or in full).

Step 1: Plan Your Visits

Level 2 Main Process: Planning Your Visit

1 — Develop the Charter

At this point, you should have organized your VOC Team, which is a small number of individuals who are responsible for the VOC.

Key questions:

- What is the objective of our visits?
- Why contact?
- When to contact?

See: Charter Tool

2 — Prepare Sample Selection

Key questions:

- Who to contact?
- Where to contact?
- How many contacts?

See: Sample Selection

Important: Once you have completed your sample selection you can initiate identification of the actual customers and coordination of the visits. Remember, people's calendars are very tight — it may take several weeks just to coordinate the dates.

3 — Develop Your Discussion Guide

Key questions:

- What information is needed?
- What information is not needed?
- How will we ask for the information?

See: Discussion Guide

Important: Selection of the team can happen prior to this step.

4 — Select Interview Team(s)

Key questions:

- Who will make the contacts?
- Are they qualified to make the contacts?
- Do they have the time to make the contacts?

Where You're At

At this point you and your team have finished the planning phase and are ready to begin making customer visits and conducting customer interviews.

See Also: Customer Contact Planner Tool

See Also: Charter Tool

See Also: Sample Selection

See Also: Discussion Guide

Step 1: Plan Your Visits

Level 2 Main Process: Customer Visits Cont'd

CONT'D

5

CONDUCT THE INTERVIEWS

Key questions:

- Are the logistics in place?
- Have customers been notified?
- Have you informed your sales force?

6

CAPTURE THE INFORMATION DURING THE INTERVIEW

TIPS ON LISTENING TO YOUR CUSTOMERS

- Sit at an angle from your customer, as it is less threatening.
- Show genuine interest such as eye contact and affirmative body motions. Avoid staring at the ceiling or checking out.
- Do not avoid silence. It's necessary.
- Make the speaker feel heard by acknowledging what was said. Clarify and verify to make sure you have understood what was said.
- Probe until you get the full story. Use of the question "Why?"
- Concentrate to the point of exhaustion!

TIPS ON TAKING NOTES

Note-taking is essential to develop high competency in capturing the messages.

- Allow the customer to talk naturally; don't force any artificial framework beyond the discussion guide. The information from the customer will be non-linear and seemingly random.
- Write in first person. Don't write "he said" or "she said."
- Take notes verbatim Don't edit the customer's words. Don't interpret what you hear. Processing of words during an interview injects your own biases. Do verify and confirm.
- Do not react to what is being said. Body language from you will bias the customer.
- Audio and/or video record the interview.

Important

It is best if you are able to audiotape the conversation (90 percent of customers will accept); that way you can listen to the customer at your leisure and also transcribe more accurately. If at all possible, take photographs or even video, with permission from your customer.

7

DEBRIEF IMMEDIATELY AFTER EACH VISIT

Key questions:

- Has the data been consolidated?
- Have the trip reports been started?
- Have key points from the visits been identified?

As soon as you complete the interview, take some time to review, clarify, and correct your notes. Complete any missing information. Add any details you may have missed.

Because you do many of these interviews, it is difficult to remember who said what; therefore, taking time after every interview will dramatically improve the quality of your notes.

This is not the time to interpret. The focus is on clarity and documentation for later use.

8

PREPARE FOR THE NEXT VISIT

Key questions:

- What do we need to do differently at the next visit?
- Did the discussion guide work?
- Did we keep our roles in mind during the visit?
- Did we archive the notes from the previous visit in a safe place?

Repeat Steps 5-8 for each customer visit.

See Also
Post-Visit Process
For next steps

Step 1: Plan Your Visits

The Charter for a Set of Customer Visits

Introduction

The first step in planning a successful set of customer contacts is to define and document the reason for the visits. Use a template like the Charter Tool, which is discussed in detail later in this section. Like any other project, you always start with the objectives, timeline and budget — a set of visits is no different.

The Charter Tool is a way to collaborate and document the reasons, intent and other relevant information for a set of customer visits. The purpose of this tool is to stimulate the thought process, collaboration and sharing of information. Most people discount this step at their expense, only to recognize later that they didn't have the same ideas about what needed to be done and why,

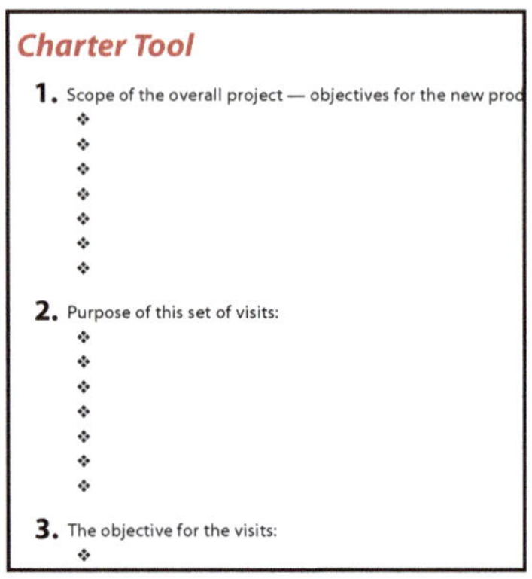
Charter Tool

1. Scope of the overall project — objectives for the new prod
2. Purpose of this set of visits:
3. The objective for the visits:

Charter Tool

General Rules for the Charter Tool

- Much of the information may already be available in other documents — for example, your project proposal, the business plan and the engineering development plan.
- It is best to develop the Charter Tool in collaboration with the members of your team. An alternative is to form a sub-team or a Core Team to complete the form and then share it with the group.
- While some of the information may seem trivial to you, it is likely that some of the members of your development team will appreciate the information. They may not be as familiar with the project as you are.
- Another use for this tool is to validate assumptions with your management chain and other key stakeholders. We suggest you schedule a meeting to discuss the form, rather than sending it. This will provide full attention and immediate feedback.
- Keep the members of your team who are not based locally — for example, those in other countries or companies — engaged and informed

Important Notes

- Make this tool the first step in the process, before any customer contact is made.
- Get engineering and marketing persons to collaborate on the completion. This will help with clarity and alignment.
- Pay particular attention to the Roles and Responsibilities section in the charter form. Many misunderstandings will be avoided if you address issues at the start of the project.

Step 1: Plan Your Visits

Writing the Purpose and Objective Statements of Your Visits

Your set of customer visits has both a purpose and objectives. The purpose can be seen as the reason why you need to interact with your customer; i.e., the intent of your visits. The objectives are the check-list of quantifiable tasks, or action items, that you use to determine if you have achieved your intent.

- Make it a collaborative effort, which will ensure synergy and, most of all, early adoption. Your team members have their own needs and are most likely to shed light on the important components of your objectives.
- Keep management informed — they hold the purse strings and can also be strong supporters of your efforts, provided you keep them engaged and well-informed.
- Customer visits deal in the qualitative space; they are not to quantify, forecast, market-size forecast growth rates, measure, select or track.

Purpose

The Purpose Statement captures the intent for your set of customer interviews. It is the articulation of the mission of the set of interviews; some people call it the goal. VOC is qualitative research; consequently, it is critical that you and the rest of your team have clarity as to the intent of your interviews, since it is not the role of the customer visits to quantify or extract quantifiable data from interviews.

The protocol for articulating a Purpose Statement is not to use words that are quantitative (measurable) but rather words that capture your overall desire. Purpose Statements have two parts: the "to [verb]..." purpose and the "in order to..." objective. "To" is followed by an verb such as "learn" or "explore" (see the table on the next page for more examples). The "in order to" statement captures the information you are trying to obtain at the end of all your customers interviews; that is, your desired outcome of your visits.

Objectives

Objectives are quantifiable metrics to ensure that your set of customer interviews do not falter or become chaotic. Generally, the objective should list the number of interviews that you intend to complete, a timeline with critical dates and clear segmentation or description of the sample selection. Naturally, any other information that might improve clarity for you, your team and your executives can also be included.

Examples of Objectives

- By December 15, 20XX, complete a set of six visits, with a total of 12 interviews with assemblypeople in the U.S. market. Write a report, present to the team and incorporate the VOC in the next design of "left-handed machines."
- For the "star" project we will conduct 15 customer interviews in China prior to the start of the definition stage. We will write a report, present it to the team and use it to define the major attributes of the "star" product line.
- By the end of the definition stage, we will initiate a series of 12 customer interviews to validate the set of "wow" features included in the new product. We will write a report and present it to the development team in time to make changes to the final design.

Step 1: Plan Your Visits

Typical Purpose Statements and Objectives

"To…" Statement Purpose Word	"In Order To…" Statement Desired Outcome of Customer Visit
Learn	To learn the buying habits of design engineers when selecting high-end test equipment in order to determine their decision making process and purchase priorities.
Investigate	To investigate the barriers and frustrations the end-users have when unpacking our products in order to document ease of use improvements to the unpacking and set up experience.
Probe	To probe the critical purchase parameters used by our end users when selecting a new product in order to capture buying process and its critical criteria.
Generate	To generate a list of the most common complaints have about our current products in order to improve the user experience in the next generation of products.
Explore	To explore the ways our end users use our equipment in order to document the "use case" frustrations and barriers so that we can improve the ease of use.
Inquire	To inquire about the likes and dislikes of senior mechanical engineers in small manufacturing companies in order to create a usable "persona", which our design team can apply when making design decisions.
Understand	To understand the critical success factors that directors of chemistry laboratories use to set priorities for their labs in order to determine opportunities for our new generation of lab products.
Describe	Describe the process our customer use to perform the set of tests to qualify their designs in order to improve the ease of use of our products.
Ascertain	To ascertain the value of technical support for software engineers who use our type of applications in order to improve our set of tech support offerings.
Identify	Identify the decision making process in mid-size manufacturing companies for our type of products in order to determine the value proposition for all decision makers and influencers.
Delineate	Delineate the way that large mining companies prioritize the purchase of new equipment for copper mines in order to document the "compelling reasons to buy" and their priorities.
Profile	Profile the engineering departments of large automotive companies in the U.S. in order to document organizational structures, organizational development and size.
Discover	Discover ease of use opportunities when customers install our product in their air conditioning systems in order to improve ease of installation opportunities and other productivity improvement of the installation process..
Study	To study the top use cases of "incoming inspection" in order to optimize our product to reduce "inspection time."

Step 1: Plan Your Visits

Charter Tool Types of Questions

This list of questions illustrates the type or nature of the questions you might want to address when developing your charter document. The list is not a survey, and is for illustration only; you need to adapt it to your business objectives and the culture of your organization.

1. Scope of the overall project — objectives for the new product or application:

For example:

- What is the new product intended to do in the market?
- How should customers perceive this product when it is announced?
- What are the objectives of this new product?
- Is this a replacement of a previous product or product line?
- Should the new product be revolutionary?
- Is the intent of the new product to reduce the manufacturing costs?
- Is the new product aimed at a new market?
- Is the new product truly new — that is, there is nothing like it in any of the addressed markets?

2. Purpose of this set of visits:

For example:

- What is the business reason for this set of visits? That is, what is our company's motivation to develop this product?
- Why are we doing this set of visits?
- What is the mission of the team doing the visits?
- What will these visits *not* cover?

Note: Use words such as discover, understand, explore, observe, internalize, learn.

3. The objective for the visits:

For example:

- What is the verifiable indicator that will show that we have met the objectives of our customer visits?
- What do we expect to have, once the visits have been completed?

Notes: Use date and time windows. Identify the number of visits and the number of team members.

4. Scope for the Voice of the Customer in this project (see purpose and objective above):

For example:

- What specific aspects of the new products will be addressed by this set of visits? For example, the GUI, the ease of use, the technical performance, objections, fit-for-use, safety, etc.
- What are the specific areas of study for this set of visits?
- What are the specific areas that will *not* be addressed by this set of visits?
- Is there a geographic set of boundaries?

Important: This form ought to be used at the very beginning of defining, deciding, and planning a set of customer visits or contacts. It can be used for face-to-face, telephone, and other types of venues for customer contact.

Important: Items 1-3 are the most critical elements of this form. Ensure that you develop these items in collaboration with your team and managers. Allocate enough time for dialog, and reflection — this process may take hours, several calendar days even, to complete. You need to align the team on common objectives.

Important: Following this template, there is a form with the types of questions you need to answer for each of the items in your charter. There is also a fully developed example.

Step 1: Plan Your Visits

5. Key dates for the set of visits:

For example:

- ❖ The official start date.
- ❖ The expected completion date.
- ❖ Don't forget key milestones. For example: when the discussion guide will be ready, when the sample selection will be completed, when the final report will be completed, when the first trip will take place.

6. Who leads this set of customer visits?

For example:

- ❖ The one person who is responsible and accountable for the success of this set of customer visits.

Note: This is the person who has the time, skills and abilities to lead the team and manage the process to ensure success.

7. Who will pay for this set of customer visits?

For example:

- ❖ What or whose budget will pay for the costs?

Note: lodging, travel, ground transportation, etc.

8. Who will approve or disapprove the Marketing Requirements Document (MRD)?

For example:

- ❖ Who is the decision maker ?
- ❖ Where does the buck stop?

9. Team members for this set of visits:

For example:

- ❖ Who are the individuals who make up the visit team?
- ❖ Who is going on the actual visits?
- ❖ Who will process the information?
- ❖ Who will generate the final report?
- ❖ Who will present the final report for approval?

10. Sponsorship:

For example:

- ❖ Who is the senior executive who will help ensure success?
- ❖ Who will mentor the team?
- ❖ Who can the team turn to for assistance and counsel?

Step 1: Plan Your Visits

Example: Charter Tool

1. Scope of the overall project — objectives for the new product or application:

For Example

- ❖ To define, develop and introduce to the market by the ABC Trade Show in April 20XX, the first product of a brand-new platform to enable design engineers to save time in trouble shooting new designs by increasing the productivity of existing modeling tools. The new product is an extension of our current portfolio to our existing customers, and does not replace any previous products.

2. Purpose of this set of visits:

- ❖ To understand and document the key requirements for the new platform (platform attributes). In this set of contacts we will not discuss the features of the first product of the platform, but rather the overall attributes of the new platform.

3. The objective for the visits:

- ❖ Starting March 1, 20XX, conduct six interviews in the U.S., six in China, and three in Europe. Develop, document, and prioritize a set of platform requirements to enable engineering to incorporate the requirements into the platform. The budget for this effort is contained in the overall development funding.

4. Key dates for the set of visits:

- ❖ The official start of the entire program is January 2, 20XX.
- ❖ The start of the customer interviews is March 1, 20XX.
- ❖ Engineering expects the documented and prioritized set of platform requirements by June 1, 20XX.

5. Who leads this set of customer visits?

- ❖ The Product Line Manager (Mary Smith) will be responsible for the execution of the purpose and objectives of this program.
- ❖ The Voice of the Customer Team will be made up of the Product Line Manager (Mary Smith), the Hardware Engineering Manager (Joe Rogers), the Software Engineering Manager (Julia Child) and the Usability Manager (Freda Wolfgang).

6. Who will pay for this set of customer visits?

- ❖ The budget is contained within the funding for the new platform.

7. Who will approve or disapprove the Marketing Requirements Document (MRD)?

- ❖ The General Manager (Sarah Rue) will be the final word; the Voice of the Customer Team will make the final recommendations and collaborate with the General Manager.

Step 1: Plan Your Visits

Methods of Observations

Introduction

Your objectives will determine the type of interaction you have with your customers. The good news is that many approaches are available.

> **Essential**
> "Ethnography is all about observing customers in their environments but without interrupting them. It can be mapped in between the upper and lower right quadrants."

The diagram below shows all forms of customer interaction, and may help you determine the type of contact you need to achieve your objectives. The diagram maps the types of interaction based on the customer's environment and the degree of interaction. None is better than any other — it depends on your objectives. For customer visits, the upper right quadrant is best. The diagram below may help you determine the type of contact you need to achieve your objectives.

The method used in this guidebook belongs in the upper right quadrant, shown in green. Incidentally, the lower right quadrant is all about ethnography; i.e., observing customers in their environments but without disrupting them.

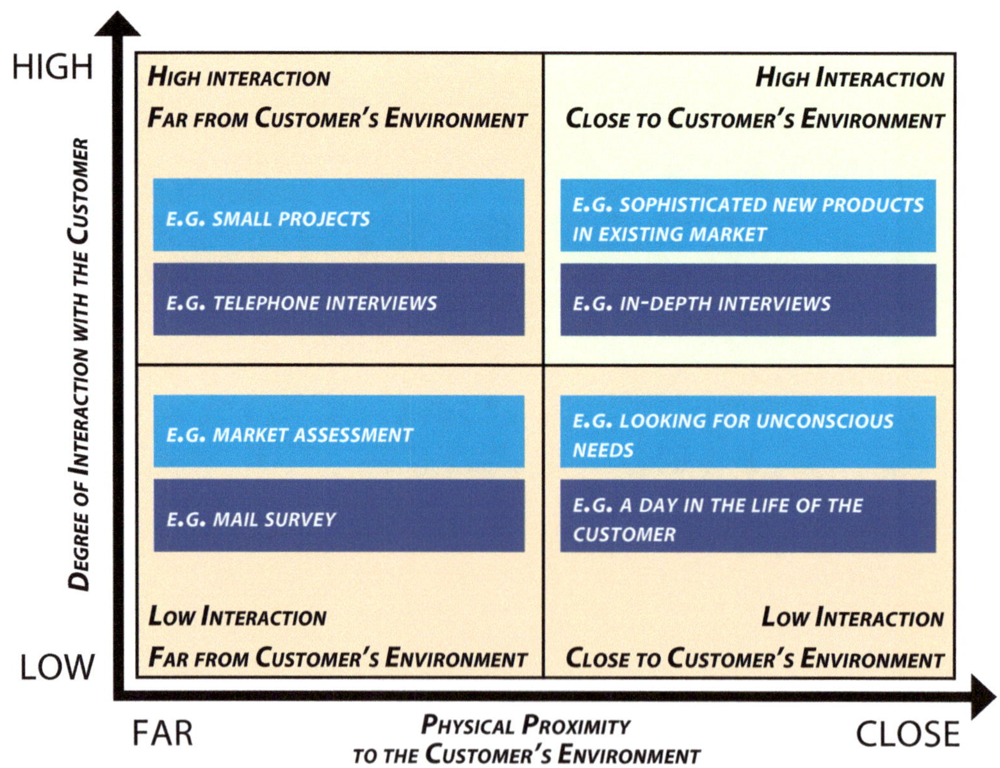

All Forms of Customer Interaction — The focus of this guidebook is on the skills and knowledge of the upper right (green) quadrant.

Personas

Profile a Typical Customer

A persona is a customer archetype that can be used by the product team to represent the typical customer. A persona is a set of user characteristics that include behavioral patterns, personal and company goals, skills, attitudes, surrounding environment, and other details. Personas are a way to profile a market segment or a group of customers by describing a single person that aggregates the common traits of the rest. Personas are a way to make an abstract market segment come to life by describing a human who possesses the common characteristics of the segment.

Teams that use personas to the fullest give them names and real personalities. These teams use all the communication tools available to them to make the personas a vivid image in the minds of the team members. Team members then use the names of the personas in discussion, and everybody can relate to the represented people and their problems.

> **Definition — Persona**
> A customer archetype that can be used by the product team to represent the typical customer.

> **Definition — Ethnography**
> A way to understand customer behavior through observation rather than in context, using several complementary methods that always include observation and some level of interviews. The value of this approach is to reveal needs that people are not conscious of or cannot articulate.

Why Personas

- They provide a personal perspective that allows for a common point your team can see.
- They set a boundary by keeping other customer groups from getting in the way of your conversation with your team.
- They help prioritize features by focusing on a single entity.
- Personas allow your team to stand in the shoes of your customers.
- Personas are a way to compel your team to be user-centric.

Not all development projects require the use of a persona. It takes time and effort to develop one, so save it for the most strategic projects — for example, when addressing a new market or when developing an entire new platform. The good news is that once you have developed them, personas can be reused for other development programs. Also, once you develop a persona, keeping it up to date does not take much effort.

Developing a Persona

Naturally, the only way to describe a valid persona is to have visited a reasonable sample of the targeted customers. The persona is a branch of ethnography, which is a way to identify customer behavior through observation rather than questionnaires. The focus is on what the customers do, rather than what they tell you they do. The profile of your persona is created from a series of interviews and observation sessions with actual customers who represent the market segment or group of customers you are targeting.

A persona represents behavior patterns, not necessarily job titles, and is better if you describe the flow of someone's day. Some call this a summary of "a day in the life of our customer." It is not the job description, because that rarely matches what the person really does during the work day.

After reading the description of a persona, you should feel as if you know that person completely in his job, personal life, family life, etc. (if all are relevant). A good persona provides a vivid and real description of your typical customer.

Step 1: Plan Your Visits

Level 3 Detailed Process: Developing a Persona

1

Do the Proper Research

A persona is critical to your design and must not be taken lightly.

This step requires analysis, research, and thoroughness. For example:

- Interview customers.
- Do a proper market segmentation.
- Prioritize your key customers.

Also avail yourself of current market research and use it to help define your persona. Talk to your distribution channels, and ask them questions about the typical customer. Read material in the trade press, which may provide additional insights into the persona.

Important: Developing a persona requires due diligence; i.e., the proper research. Avoid an ad-hoc approach. You must dedicate time and effort to this endeavor.

2

Describe the Personal Traits

Describe the personal traits of your persona. This is a list of the attributes that will clearly profile your persona. Ensure that your description is unique, not a one-size-fits-all. Each persona has to be clearly different from others.

Include your team in this step. This is a collaborative effort, plus the combined knowledge of your team will be very valuable.

See Also: Persona Examples

3

Document Your Persona

Now it's time to fill in the blanks with the research information.

Keep the description of your persona to a manageable size, perhaps a few pages. Avoid long documents.

Don't include details that add little or no value. Stick to relevant characteristics.

Always include what your persona trying to do or accomplish.

Involve your team in this step.

4

Develop Your Communication Plan

Define who will use the persona, and set about developing a plan for clear communication; for example, call a meeting to introduce your persona, or write a profile on your persona and distribute it to all your team members. Feel free to add graphics to your description.

5

Develop the Protocol

Write a set of instructions on how to use your persona. For example, use personas to develop cases and scenarios. Provide guidance to your team so they get the maximum benefit from using the persona.

6

Use and Improve Your Persona

Start using your persona, and at the same time start improving its quality by creating flyers, posters and other handouts to help your team. Improve the profile with new information as you get it. Keep "training" your team on the use of personas.

Step 1: Plan Your Visits

Critical Ingredients

For a persona to work, several additional ingredients must be put in place.

- **CREDIBILITY:** A persona must be based on solid research. Reasonable care in segmenting your market is the first step, followed by research with actual customers who provide the needed information. Lastly, you must amalgamate the information to create a character who is believable and truly representative.
- **COMMUNICATION:** Once you have a persona, you must proactively share it with the rest of your team, including engineering, documentation, and others. It is not enough to create a persona — you must deploy it in the organization so others can start using it.
- **PROTOCOL:** You must also provide guidelines as to how your team will use your persona. A set of instructions will be very helpful, together with a meeting where you describe the use of a persona.
- **PACKAGING:** Your persona is not a replacement for other forms of market research. It is an enhancement to a package of quantitative and qualitative research, which together provide the needed input to develop successful products.
- **PERSEVERANCE:** If your team has not used personas before, you will have to persist and demonstrate the value of using them. At first, the use of personas may not be comfortable to some members of your team. Refer to the Recommended Books section of this guidebook for articles and books on how others have used this powerful tool. Additionally, do a Web search to obtain articles, books and other materials that reinforce the benefits of using personas in product development.

Questions to Ask about Your Persona

These questions are intended to help you describe your persona. Not every question must be answered. Generate additional questions that are more relevant to you and your team. Use your judgment by keeping your objectives in mind.

See Also
Recommended Books

Personal Questions

Only include personal information that is relevant to your product or objectives — just enough to make it real.

1. What is his name?
2. What is his title?
3. How long has he been in his current job?
4. What company does he work for?
5. What is his environment like?
6. What tools does he use?

Advanced Questions

1. Why does he do what he does? (This is the most critical question for your persona.)
2. What are his behavior patterns?
3. What problem is he trying to solve?
4. What results is he striving for?
5. What are his goals for his job?
6. What skills does he possess?
7. What type of information does he require to perform his duties?

Step 1: Plan Your Visits

8. What are his job attitudes?
9. How is his performance measured?
10. What is are stressors about his job?
11. What does he fear?
12. What is his job description.
13. What are the most typical tasks that he performs in his job
14. What are his surroundings like?
15. What company does he work for?
16. What region does he represent? What area of the country?
17. Does he have an equivalent in another country? And what are the key differences?
18. What market segment does he represent?

Questions about Your Product

The word product is used generically to indicated the value you are trying to sell, which could be an existing product, a new technology , a service, etc.

1. Why does he use your product?
2. When does he use your product?
3. What is he trying to accomplish when he uses your product?
4. What problem or challenge does he have that makes him use your product?
5. What product does he use now?
6. What happens before he uses your product?
7. What happens after he uses your product?
8. What benefit does your product provide him?
9. What are the joys he gets from using your product?
10. What are the frustrations he has when using your product?

FAQs about Personas

Q. How Many do I Need?

A. The rule of thumb is three or fewer; the more you have the more difficult it will be for your team to tell them apart. The need for multiple personas depends on how different the market segments or groups of customers are in relation to each other. For example, you might describe two personas in a market that has older and younger workers. You might want to use personas to describe the hierarchy of the organization, such as manager, supervisor and individual contributor.

Q. Are Personas the Same as Market Segments?

A. Generally not. Your personas may emerge initially from your existing market segmentation, but their behavior patterns will create new labels — for example, "the corner-cutter" for someone who skips steps or "the detail follower" for someone who thrives on the minutiae of his job.

Example: Mapping a Persona

Persona for Customer Experience Engineers in Software Companies		
Name and Title	A name makes the persona feel real. A title links the persona to a market segment	Sarah, Chief Customer Experience Designer.
Education and experience	Describe the typical education of your target persona, include higher education degrees, certificates and other training.	She is a degreed engineer in electrical engineering, with some graduate work. She has attended seminars on usability, ethnography and market research.
Work activities and environment	Look at your persona's job description, reporting relationships, influences, role, goals and objectives at work	Sarah is responsible for developing and implementing all ease of use tests in her organization. She reports to the VP of Engineering. Sarah has much influence in the purchase of our equipment for testing ease of use. She is also gaining much respect with her peers as ease of use becomes more important in the purchase of our products.
Process	Identify where and how your persona fits into the process on the job, both formally and informally.	She works closely with the marketing organization, but her primary audiences are the mechanical and software engineers as they define the ease of use for their products.
Goals, fears and aspirations	Identify job and career related concerns.	Not many people start as customer experience professionals, as it is a relatively new career. Sarah started as an EE and eventually acquired the additional skills needed to do the job. Most of her peers also have similar career paths.
Skills, knowledge and abilities	In addition to education, investigate the your persona's abilities and required job skills.	Sarah is very skillful in empathizing with the people who buy their products. She is committed to her field, and continues to attend seminars and read books on the topic.
Attitudes	Discover your persona's attitudes toward people, the job, vendors, technology, etc.	Sarah loves her new career as a Chief Customer Experience Designer. She can combine her EE training with her passion for technology and customer experience.
Impact and influence	Clearly understand your persona's role in the purchase, use and replacement of your product.	She can also influence the entire engineering effort and is gaining more influence due to the importance of ease of use.
Additional references	List information contained in other venues; for example, photographs, diagrams and any other relevant details.	Photos taken during the observation visit to Sarah's office; see Fred H. for the album. There is also a job description on our intranet.

This example is a starting point; it does not yet constitute a full persona. View it as the first pass description, which will grow over time as you discover additional information about your persona. Adapt this example to best serve your business objectives and the culture of your team.

The goal is to describe a typical customer through a persona such that it becomes a vivid description of the group of customers you have selected. It must make the typical customer real and clear to everyone in your team, even those who have not talked to customer.

Step 2: Organize Your Visits

NEW TOPIC

Sample Selection

Identify the Right Person to Visit

Sample selection is the process used to identify the appropriate number and types of customers to interview. It is clearly the most critical step in securing the voice of the customer. Not only does talking to the wrong people not help you, it may even have a negative impact by providing you with distorted input.

Think of sample selection as a way to profile the most representative spokespersons for a market segment. This profile need not be the statistical median or average. It does need to be the most qualified customer to ask the questions you have in your discussion guide and satisfy the objectives you set for the series of contacts. Sample selection is a way to ensure that the message from your customers truly represents the voice of all the customers in the segment or market.

Why Sample Selection is Critical

- Talking to the wrong people will not help you — it may even distort your results
- Sample selection is a disciplined way to visualize your markets and customers within those markets, and it's a fine way to engage your team in understanding who defines the value for your new product. Simply said, it's a way to make the customer real to your development team.
- It ensures that the input you receive is high-quality, from people who represent your typical customer.
- It makes the information you gather defensible — it is not anecdotal or accidental, but a controlled and level-headed process.

The Four Critical Questions

Every time you define your sample selection, you must answer four critical questions. Without the four answers, you risk talking to the wrong people in the wrong place. The answers to the four questions are solely driven by the purpose and the objectives of your set of interviews, which by now you will have finalized and documented.

1. **WHO ARE YOU GOING TO INTERVIEW?** You start by identifying the profile or function of the persons you need to interview. For example, Engineering Managers of Design Organizations. Ultimately, you must identify them by name and title. For example, Sarah, Chief Customer Experience Designer. This provides a clear direction.
2. **HOW MANY WILL YOU INTERVIEW?** Now you need to decide how many persons of each profile you will need to interview in order to achieve your objectives. Determine the number of visits based on the number of different market segments. For example, if you will visit customers in the U.S. and Europe, you will need to create two matrices. Variables such as the level of competition or economic conditions will modify your sample selection. There is no fixed formula!
3. **WHICH VENUE WILL YOU USE?** The traditional venue is the customer's office. But that may not be the most appropriate place. See Typology of Customer Visits in this guidebook for ideas of alternative venues.
4. **WHAT IS THE CROSS-SECTION?** Now you need to define what cross-section of your market segment is appropriate. For example, "small, medium and large companies," or "Europe, Asia and America" or "telecommunications, semi-conductor and power management." In short, determine how many people from each of the major groups within your market segmentation will you visit.

Warning About Sample Selection Time Frames

Obtaining volunteers to participate in a series of interviews is not difficult, but it can be time consuming, particularly if your volunteers are in other countries. You should allocate three to four calendar weeks to ensure that you have time to recruit the right persons. You first need to define your sample selection, then you need to

Step 2: Organize Your Visits

Example: The Four Critical Questions

Template

Mapping a Set of Customer Interviews.

Question	Examples			
Who Profiles (functions) of all interviewees, ultimately names and titles.	Engineering Managers	Engineering Team Leads	VP of Engineering	Director of Engineering
How Many Number of interviews for each of profile type.	3	4	3	3
Venue Forum you have selected to conduct the interviews.	Trade show meeting	Trade show meeting	Their -offices	At the next meeting of the IEEE in New York
Cross-Section Segments or groups from the chosen market.	Three from small manufacturing	Two from small manufacturing companies Two from small software companies	One from small manufacturing company One from medium-sized, software company One from multinational company	Three members of the IEEE software division

This table depicts one way to ensure that you've covered all the critical variables for a set of customer interviews by addressing the four critical questions. This guidebook includes other templates for mapping customer visits.

In this example, we have mapped 13 interviews, which is on the high end in terms of the number of visits.

Definition

Functionalize

Used in sample selection during the process of identifying and describing the function of an individual. For example, "engineering" is the function, whereas "engineering manager" identifies a specific individual.

Step 2: Organize Your Visits

start contacting potential volunteers and finally you need to schedule and coordinate the interviews. To save time, you and your team need to prepare and collaborate in obtaining volunteers.

General Rules for Sample Selection

1. Sample selection is both art and science. You must use you judgment, as there are no exact formulas.
2. Sample selection must always be driven by the objectives you have set for the customer interaction you are about to undertake. Ensure that you have clear and documented objectives for your customer visits.
3. The natural question for sample selection is, "What is the profile of the person that can give us the best quality input, given the set of objectives for our customer visits?"
4. Sample selection is not exactly market segmentation, although they are closely related. Sample selection is a one-time event to profile the best candidates who can answer your questions and satisfy the objectives of your customer visits. Market segmentation, on the other hand, is a framework that describes a market or a large group of customers; it is generally created by a marketing organization and is multi-use. You can start with the market segmentation to develop your sample selection.
5. Keep your sample selection very simple. Avoid n-dimensional tables; they rarely help. A two-dimensional scheme is best. If your market is very complicated, consider generating several two-dimensional schemes, where the aggregate depicts the full sample selection.
6. Keep it feasible. It's not necessary to visit every customer in every segment. Your objectives should narrow the sample considerably and keep you from straying into needless tangents.
7. The question always arises, "How long should each meeting with the customer be?" In jest, as long as you can get away with. In reality, 90 minutes is a good rule of thumb. First, your customers may not give you any more time, Second, if you are equipped with a good discussion guide, it should not take longer to get valid information. It is assumed that no more than three topics of conversation can be covered in 90 minutes. More than three, and you start sacrificing the quality of the information.
8. Beware of the moving-target trap. While you are making your sample selection, someone in your team suggests additional topics for discussion, which confuses the selection. Keep your objectives in mind and avoid accidental modifications.
9. These types of interviews are in the qualitative domain, and statistical sampling may lead you astray. You are selecting a few people who are representative of a customer group; that is, they have the profile to speak on behalf of a larger group of customers.

Step 2: Organize Your Visits

Templates for Sample Selection

Because sample selection is so critical to capturing the voice of your customer, the rest of this chapter focuses on four different approaches to selecting customers for interviews— Customer Archetypes, The Decision-Making Tree, The Classic Tree Structure and Market Segmentation.

No one of the templates is more important than the others. Each simply provides a different approach. Your objectives should drive which template you use. Use the templates as they are, modify them, find additional templates from other sources or create your own to better suit your objectives and style.

The key message is to use a methodical approach — sample selection is not an ad-hoc effort!

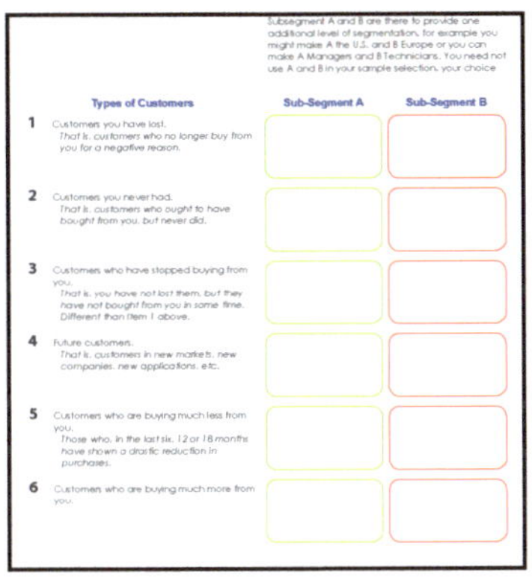

Customer Archetypes

CUSTOMER ARCHETYPES allow you to select based on various criteria to define the nature of your customers. Use them to address customers with whom you don't normally talk. This template is also called the Breakthrough Model, as it allows you to view customers in a different light.

THE DECISION-MAKING TREE is generally used to determine how the purchase decision is made. Each level has a set of needs well beyond the specs of the product. It can also be used to map the various users of your product.

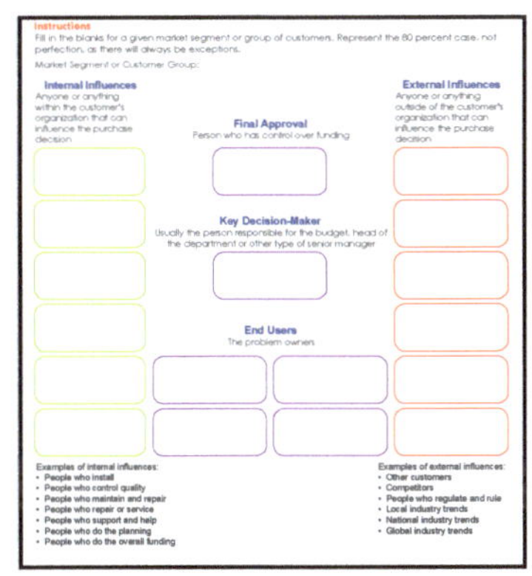

Decision-Making Tree

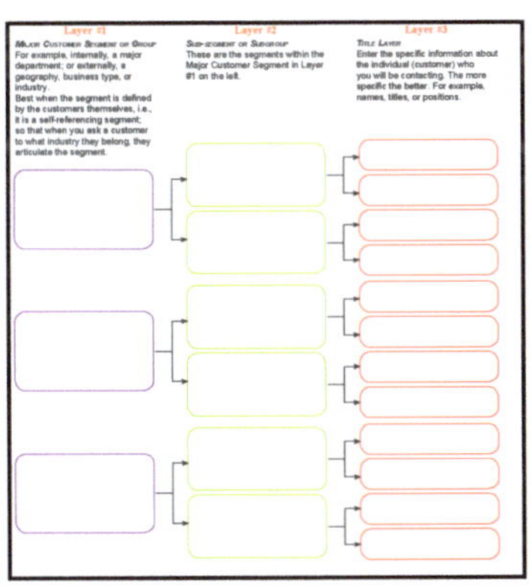

Classic Tree Structure

THE CLASSIC TREE STRUCTURE is a practical way to get to the end user. It is a good way to define hierarchy. This model also lends itself to mapping several market segments at the same time.

MARKET SEGMENTATION includes the traditional segmentation criteria, but you can use it to define your sample.

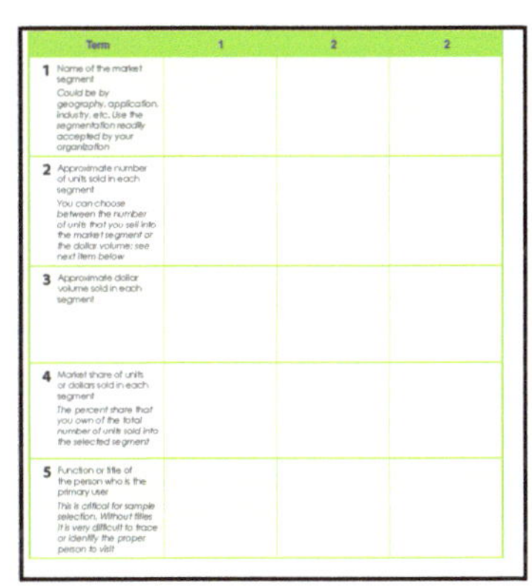

Market Segmentation

Step 2: Organize Your Visits

Example: Customer Archetypes

Subsegments A and B are there to provide one additional level of segmentation. For example, you might make A the U.S. and B Europe or you can make A Managers and B Technicians. You need not use both A and B in your sample selection; it's your choice.

Types of Customers	Subsegment A	Subsegment B
1. Customers you have lost. *That is, customers who no longer buy from you for a negative reason.*	We lost the Acme account last year. We had steady business averaging $50,000 per month. Then suddenly stopped buying.	In Europe we lost the ABC account due to an unresolved reliability issue. Customer decided to stop buying from us
2. Customers you never had. *That is, customers who ought to have bought from you, but never did.*		
3. Customers who have stopped buying from you. *That is, you have not lost them, but they have not bought from you in some time. Different than Item 1 above.*		
4. Future customers. *That is, customers in new markets, new companies, new applications, etc.*	The cell phone market will need to model their networks in 3D, which is one of our "core competencies."	
5. Customers who are buying much less from you. *Those who, in the last six, 12 or 18 months have shown a drastic reduction in purchases.*		
6. Customers who are buying much more from you.		We are seeing a sudden increase in orders from the aerospace industry in Europe. We don't know why.

Example: Customer Archetypes (Cont'd)

Using the template format of the previous page, here is an extensive list of Customer Archetypes:

1. Customers you have lost.

   That is, customers who no longer buy from you for a negative reason.

2. Customers you never had.

   That is, customers who ought to have bought from you, but never did.

3. Customers who have stopped buying from you.

   That is, you have not lost them, but they have not bought from you in some time. Different than Item 1 above.

4. Future customers.

   That is, customers in new markets, new companies, new applications, etc.

5. Customers who are buying much less from you.

   Those who, in the last six, 12 or 18 months have shown a drastic reduction in purchases.

6. Customers who are buying much more from you.
7. Your competitor's customers.

   While hard to identify, the effort is worthwhile. You will get a brand-new perspective on you, your products, and your company.

8. Your most profitable customers.
9. Your least profitable customers.
10. Customers who cost you money.

    Customers who demand large discounts and expect added features and services. In some cases, you sell to them at a loss, even though that is not your intention.

11. Upset customers.

    Customers who are upset at you but still buy from you.

12. Power users.

    Those customers who are always ahead of the curve. They enjoy pushing the limits of technology. Sometimes called the early adopters; not the same as lead customers. They like to have the latest things, even though they may not need them.

13. Lead customers.

    Customers who are developing new products that may revolutionize a market. Those that are creating breakthrough products. Not the same as power users.

14. Channel partners.

    Your distributors, VARs, REPs and other alternate channels of distribution. Naturally, also their customers.

15. Lost leads (almost customers).

    A customer who expressed an interest in buying your product, but either did not buy or purchased a competitor's product.

16. Mix-and-match customers.

    Customers who buy from you and your competitors. Typically they like your products, but they still buy from your competitors as well.

17. Distant customers.

    Customers who are the farthest away from your office. Generally, they have a very different set of perspectives.

18. Define your own category.

Step 2: Organize Your Visits

Example: Decision-Making Tree

This approach focuses on how decisions are made in an organization. Use this approach to identify the needs of the different stakeholders; for example, if more than one person is involved in deciding the purchase of your product. This approach is particularly helpful if your company makes capital equipment that require many approvals.

When filling in the boxes, represent the 80 percent case — not perfection — as there will always be exceptions.

Market Segment or Customer Group: ______________________________

Internal Influences

Anyone or anything within the customer's organization that can influence the purchase decision.

- The Quality Manager
- The business unit Financial Manager
- The Engineering Managers who work for the VP of Engineering

Final Approval

Person who has control over funding.

- The Business Unit General Manager

Key Decision-Maker

Usually the person responsible for the budget, head of the department or other type of senior manager.

- The VP of Engineering

End Users

The problem-owners.

- Technicians Level 1
- Technicians Level 2

External Influences

Anyone or anything outside the customer's organization that can influence the purchase decision.

- Our main competitor, "Acme Electronics"
- The industry standards organization
- Government regulators

Examples of internal influences:

- People who install.
- People who control quality.
- People who maintain and repair.
- People who repair or service.
- People who support and help.
- People who do the planning.
- People who provide the overall funding.

Examples of external influences:

- Other customers.
- Competitors.
- People who regulate and rule.
- Local industry trends.
- National industry trends.
- Global industry trends.

Step 2: Organize Your Visits

Example: Classic Tree Structure

This traditional approach is easy to use. The highest level in the hierarchy is in the left column. The target person you wish to interview is in the right column.

Use this approach to identify the sample by the titles of their functions. In the end, you have identifiable titles you can call a company and ask for.

Highest Level | **Example #1** | **Target**

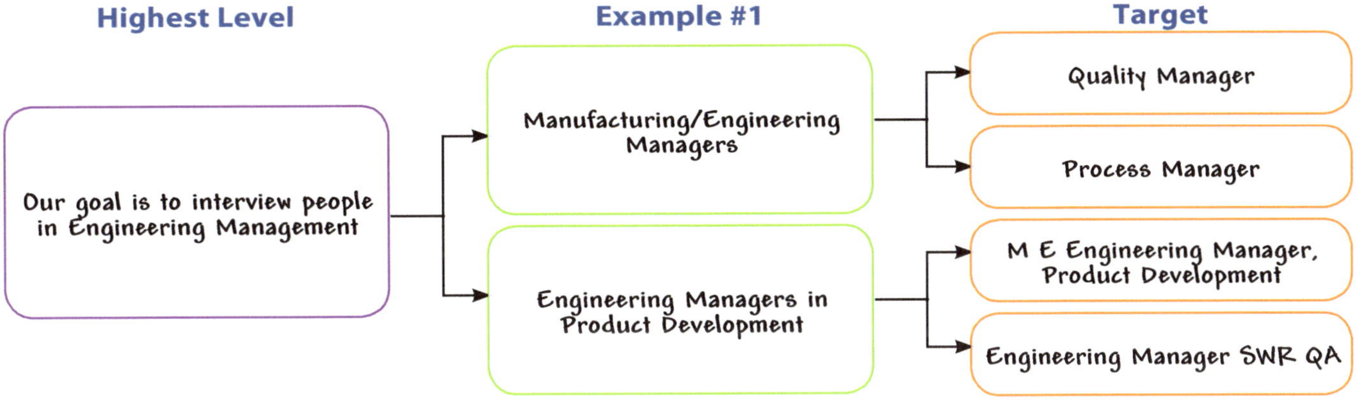

Example #2

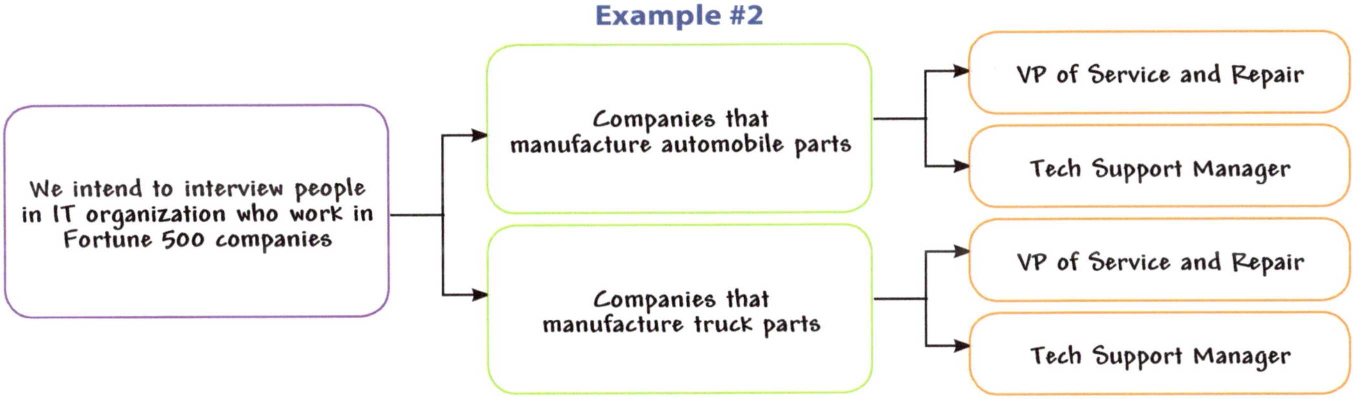

Example #3
(Consumer Product)

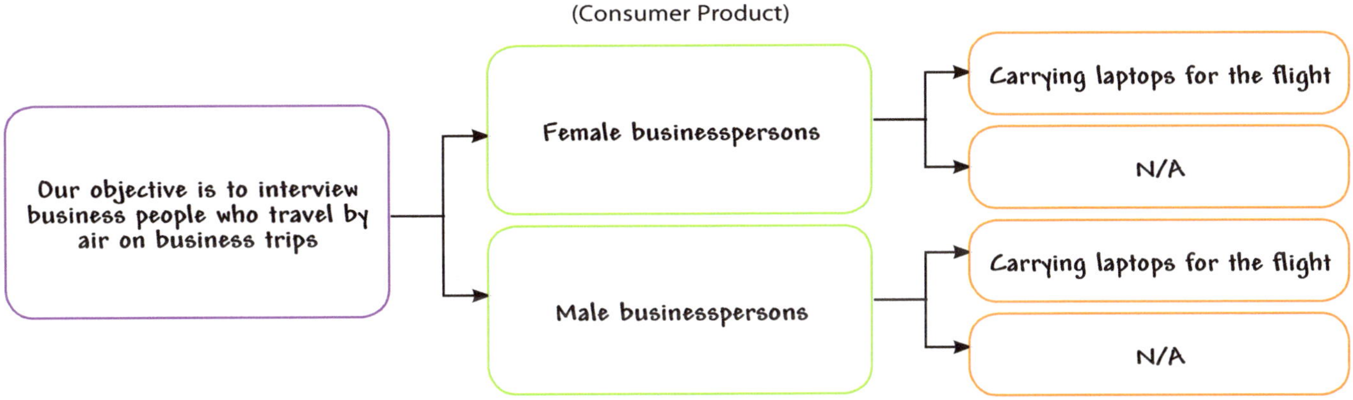

Step 2: Organize Your Visits

Example: Market Segmentation

This sample selection approach uses the traditional market segmentation model. It's important to note that you still need to identify the individuals to interview by name and title.

Term	Market #1	Market #2	Market #3
1. Name of the market segment. *Could be by geography, application, industry, etc. Use the segmentation readily accepted by your organization.*	Mechanical engineering in manufacturing companies of off-road equipment	Technical documentation in manufacturing companies of heavy-duty highway trucking	
2. Market share of units or dollars sold in each segment. *The percent share that you own of the total number of units sold into the selected segment.*	20%	30%	
3. Function or title of the person who is the primary user. *This is critical for sample selection. Without titles it is very difficult to trace or identify the proper person to visit.*	Lead Mechanical Engineer Engineering Managers, Mechanical Engineering	Mechanical Drafting Technician Manager of Technical Communications	
4. What does the primary user do with the product?	Enters information for service manuals for new trucks	Edits information in service manuals	
5. Function or title of the person who makes the purchase decision. *This is critical for sample selection. Without a title it is very difficult to trace or identify the proper person to visit.*	VP of Engineering	VP of Engineering	

Customers to Avoid

Traditional Customer

The truth is, no customer should be avoided, but when selecting your sample keep in mind your "friends and relatives" customers. These are the traditional customers who get all the visits from your company. Generally, they are on good terms, well-liked by your sales organization or distribution channel, and have not had any trouble with your company in some time.

The friends-and-relatives issue surfaces when you use your sales organization or distribution channel to select your sample. It is natural that they will suggest people who bought from them recently; they are not likely to suggest a customer that for some reason is upset with them, your company or your product.

Thus, your sales organization must be kept informed of who you visit, but they should not be the sole source of candidates. We would further declare that, in some cases, they may not be the best source for contacts. Keep your objectives in mind and the criteria you have defined for your sample selection.

Alternative Sources

There are many sources of information besides your sales organization:

Examples
1. Use the "bingo cards" magazines or telephone calls to your support organization that customers send in to inquire about your product. By definition, they are already interested in your products. The same approach can be done when they click for additional information on your Web site.
2. Your service, repair and maintenance organization is always a good source of names and titles.
3. Professional organizations can be utilized to obtain volunteers. Be certain to observe the rules and guidelines to approach members.
4. Obtain names and titles from customers who call your support center. Simply ask them if they wish to volunteer to provide input on future products.
5. Blogs, industry Web sites, and other special interest groups can be approach to solicit volunteers
6. Lists from tradeshows, expositions, and conferences are available for purchase from the organizers of the event.
7. Your trade show or exposition booth is always a good place to solicit volunteers. Post a sign asking for volunteers to provide input on future products
8. Purchase lists from List Brokers. These are companies that specialize in creating extensive lists specific to industries or other interest groups.
9. Retain a Market Research or Market Survey firms to obtain names for you. They have the capability to call many individuals in your industry and recruit volunteers
10. Use your imagination. Call a meeting of your team and brainstorm how to obtain volunteers. You will be surprised how many good ideas emerge.

Step 2: Organize Your Visits

Selecting Who You are Going to Interview

See Also Personas

You must have a clear profile of each person you will interview. Consider that your sample is always, between three to 24; thus, you will interview individuals who represent an entire class of customers. Consequently, identifying the "who" is not only critical for logistical purposes, but also for strategic intent. Do your due diligence to first profile, and then find, those you wish to interview.

We recommend using personas to profile the right targets in a very clear fashion.

Finding Customers to Interview
Use the bingo cards magazines or telephone calls to your support organization that customers send in to inquire about your products because they are already interested in your products. The same approach can be done when they click for additional information on your Web site.
Your service, repair and maintenance departments are always good sources of names and titles.
Professional organizations can be utilized to obtain volunteers. Be certain to observe the rules and guidelines to approach members.
Obtain names and titles from customers who call your support center. Simply ask them if they wish to volunteer to provide input on future products.
Blogs, industry Web sites, and other special interest groups can be approach to solicit volunteers.
Lists from trade shows, expositions and conferences are available for purchase from the organizers of the event.
Your trade show or exposition booth is always a good place to solicit volunteers. Post a sign asking for volunteers to provide input on future products
Purchase lists from list brokers. These are companies that specialize in creating extensive lists specific to industries or other interest groups.
Retain a Market Research or Market Survey firm to obtain names for you. They have the capability to call many individuals in your industry and recruit volunteers
Use your imagination. Call a meeting of your team and brainstorm how to obtain volunteers. You will be surprised how many good ideas emerge.

Be Careful of Favorite Customers

No customer should be avoided, but when selecting your sample be wary of the too frequently used "friends and family" customers, who have not had any trouble with your company in some time. It's natural that your sales organization or distribution channel will have a good rapport with these customers and, therefore, suggest them first for the sample selection; whereas, considering customers who are upset with your company or product might be overlooked.

Your sales organization or distribution channel must be kept informed of who you visit, but they should not be the sole source of candidates. We would further declare that they may not be the best source for candidates. Keep your objectives in mind and the criteria you have defined for your sample selection.

Step 2: Organize Your Visits

Selecting How Many You Will Interview

How Many is Enough

There is no formula to determine the ideal number. The standard range for visits is three to 24. More than 24, and you should probably seek a different type of tool; for example, a market survey. Fewer than three will not give you the enough data to determine trends. Also, if you wish to contact design engineers and software engineers, you ought to pick at least three visits from both groups.

Two other variables that impact sample selection are time and money. While you may desire to visit 100 customers, it may not be feasible. This is where your objectives come into play. You must develop clear criteria for deciding on your sample and then use it to reduce the bevy of contacts to a manageable number.

One Company, Multiple Sites

Many companies have several sites or branches. Should visits to two sites be counted as one visit? For example, you are visiting two sites belonging to IBM; how many visits should you do? How should you count them in the overall visit plan? Remember that you need at least three of anything to develop valid trends. Assuming that you have a valid reason for visiting two IBM sites, then you need three interviews per site.

An added complexity, assume that you need to talk to engineering managers and technicians. What is the count then? Well, you need to interview three engineering managers per site and three technicians per site; therefore, the total number of interviews for both sites is 12. In this case, you better have a valid reason for so many.

Important

Sample selection is not difficult but it can be time-consuming. Your team can help you save time recruiting volunteers.

Plan the Number of Visits

Number of Customer Contacts Based on the Type of Development

In the worksheet on the next page you can estimate the number of interviews you will need based on the type of development program (the X axis) or your approach (the Y axis), which is driven by your objectives.

You can choose to select the number of interviews based on functional titles; for this, use row number 3. Perhaps you need to contact customers from a single company who are in different sites; in this case use row number 1. Finally, you may have a combination, in which case you fill the cells based on your objectives.

Note that this worksheet is not strict; it is simply a way to help you select the right number of interviews. Further, you can use the matrix to develop your own tool, one which better meet your needs and fits better with your markets and customers.

And remember that the experts suggest that you should do no fewer than three interviews total nor more than 24. Beyond that, you are into the realm of market research.

Sample selection is art and science. You are in the area of qualitative research. You must use you judgment, as there are no exact formulas. Here are some rules of thumb from the "school of hard knocks":

- Better to have a two-hour interview with one customer than four half-hour interviews with four customers. Depth is more important than volume.
- Better to have two of you conducting the interviews than only one. Two sets of eyes and ears provide stereo vision of the customer.
- The more strategic your program, the more customers you should interview. Obviously, the consequences are larger as the strategic value of your program increases.
- Fewer than three customers is not a programmatic set of customer visits — it's not enough samples to create a pattern. The intent of customer visits is to develop messages with a pattern.

Step 2: Organize Your Visits

Example: Determining How Many Interviews to Do

Please keep in mind the VOC (Voice of the Customer) interviews are a form of qualitative research. There is no formula to calculate the right number of interviews. Nevertheless, the consensus of the experts such as Dr. Edward McQuarrie, the authors of this guidebook and others is that the limits are no fewer than three and no more than 24. The table below provide additional guidance based on the type of development you are managing.

Interviews	Type of Development Program		
	3 or more (maximum 12)	5 or more (maximum 12)	7 or more (maximum 24)
	Minor modification to an existing product — for example, an A version or a maintenance release	A new product from an existing platform	A new platform aimed at a new market dramatically different from your current products
Individuals from a single company with many sites			Interview six engineering managers from Acme Electronics — two in the U.S., two in China, and two in Germany.
Individuals to interview per each site			Interview five Engineering Team Leaders in medium size, telecommunication companies in the U. S.
Functional types — i.e., the title of the customers to be interviewed			Interview five Engineering Team Leaders in medium size, telecommunication companies in the U. S.
[Define your own category]			

In the above example, the development team will be creating a new platform (column 3) and has decided to do 16 interviews. This is very appropriate, as considerable customer input will be needed to design a new platform.

Step 2: Organize Your Visits

Selecting Which Venue You Will Use

Introduction

The natural inclination when it comes to interviewing customers is to travel to their place of business and, by and large, this is correct. On the other hand, there are many places in which to conduct very productive interviews. This section covers the many alternatives to explore; some may be more appropriate to your objectives than the traditional let's-travel-to-their-office approach.

Furthermore, there is the added dimension of the frequency of interviews. The traditional visit to their place of business is generally a "single-shot event"; that is, you do not return to that particular customer. It may be more productive for you to find a venue where there is opportunity for continuous interaction with your customers. This section also provides ideas for repeat contact.

Finally, venues other than their offices may be more conducive to open dialog. Customers are not under the pressure of meetings, interruptions, etc. Being able to get them away from their office may work in your favor.

Definition

Typology

The systematic classification of types.

Definition

Customer Visit

Used for any time you have interaction with your customers.

Definition

Customer Interview

Used when you talk to the customer as part of a programmatic set of visits.

Important

Sample selection is not difficult but it can be time-consuming. Your team can help you save time recruiting volunteers.

Typology of Customer Contacts

Introduction

Typology is study or analysis based on categories. In the following section, we have classified customer contacts by "type"; that is, by the place and format of the contact.

The typology of customer contacts is important, as it allows you to select the right venue to suit your objectives, plus it shows that there are many ways to obtain valuable customer input.

The typology in the following pages is not all-inclusive, but it provides a foundation for you to build your own categorization.

Important Note

The typology (classification) shown in the following pages focuses only on customer contacts where you are interacting with your customer directly. It does not apply to third party customer interaction including market research methods such as focus groups, executive interviews, etc.

Context

The term "customer visits" is a generic phrase that connotes the overall set of activities to contact or connect with one's customers. Thus, in the context of customer intimacy, "visit" does not necessarily mean a physical trip, it means any opportunity to connect with the customer. For example, when talking to a customer on the phone, you are visiting in the sense of understanding customer needs.

Visiting customers need not cost a lot of money; while formal visits indeed require funding, there are many other opportunities to connect with customers and to obtain valuable information.

Step 2: Organize Your Visits

Typology of Customer Contacts: Types of Venues (Forums)

Type of Visit	Description	Application	Pros & Cons
1. Programmatic *(outbound)*	A set of visits to customer sites, with a clear and documented purpose or set of objectives. Generally, between six and 24 actual visits or meetings.	To explore specific or targeted factors about the customer — for example, to discover customer needs.	👍 Best form to identify hidden needs. 👍 Best form to establish clear patterns of needs. 👍 Best for obtaining qualitative information. 👎 To be effective requires planning and considerable time.
2. Ad Hoc	Unplanned contacts with customers. Not part of a program of visits, but part of a deliberate plan to meet and talk with customers.	Get a reading of the marketplace. Identify the current situation with the customers.	👍 Ensures that every contact with a customer, even casual ones, are useful. 👍 Gets everyone in the organization involved with customers . 👎 Difficult to consolidate the information in a useful fashion.
3. Inbound	When customers travel to your office or factory.	Invite the customers to present their needs or discuss their plans. Address a specific customer requirement or need. Allow more of your team members to have customer contacts..	👍 Your customer is focused on you and your company. 👍 Allows employees who normally do not travel to meet customers. 👎 Difficult to persuade customers to spend the time.
4. Neutral Site	When you meet in a third-party location — for example, at a conference or a trade show.	Take advantage of travel and budgets. Talk to more customers than during a single visit.	👍 Good way to meet many customers in one place. 👎 Too many distractions.
5. Customer Councils *(similar to, but not the same as, User Groups)*	Regular meetings with customers, such as a user group. Also known as board of advisors or customer panels. Repeat contact under a programmatic framework.	To obtain input from a recurring set of customers — for example, executives.	👍 Greater involvement by the customers. 👍 Goes beyond contacts and into relationship-building. 👎 Time and effort on the part of customers. 👎 May be costly to your company.

Step 2: Organize Your Visits

Type of Visit	Description	Application	Pros & Cons
6. Extended On-Site	Spending more than a day at the same customer site.	Close observation of customers in their environment. Observation of the customer's processes. Data gathering.	Total immersion in the customer site. Takes time on your part.
7. Contextual Inquiry	Customers are observed while they use your product in the course of normal business. Observation in context with the actual work.	Observation of customers under a strict protocol for the observers. The intent is to observe without influencing. Observing customers in *"their natural habitat."*	Good way to improve ease of use. Good way to discover hidden needs. Good way to observe the customers without any outside influences. Takes time and money.
8. Technical Liaison Agreement *(TLA)*	A formal agreement with a customer to exchange technical information and personnel. For example, two companies agree to exchange engineers for an extended period of time.	To provide a robust channel of communication with a particular customer. To exchange proprietary information.	Excellent way to create dialog between technical personnel. Particularly useful in developing new technology. Requires a legal contract or agreement due to exposure of proprietary information.
9. Blogs *(customer intranets)*	Organize and support Web blogs focused on specific topics. Enable your customers to access an intranet within your Web site.	Exchange of information. Determine needs, and things to avoid. Develop new concepts. Learn the latest "buzz" about your industry.	Needs consistent support on your part. Cannot become stale or boring. Your competitors may access your blog, and you may not know it. A focused forum of people who care and are of interest to you.
10. University Relations	An agreement with a university or college to obtain information. It can also be done with research institutions, "think tanks," etc.	To get input on future applications. To enable academic information to influence new products.	Relatively easy to arrange. Allows for contact with professors and other research individuals who set industry trends. May not represent your customers' opinions.

Step 2: Organize Your Visits

Type of Visit	Description	Application	Pros & Cons
11. Telemarketing	The use of your existing telemarketing or service organization to contact customers. For example, offering a small incentive to customers that call your service center in exchange for a short interview.	To obtain statistically defensible data. To spot trends.	👍Relatively easy to set up. 👎May not represent the customer segment that you have targeted.
12. Counter Duty	Stand behind the counter at a service center or in a retail outlet in a B-to-B environment and be a tech support person, take orders from customers, or work with your service and repair department.	Obtain answers to simple questions.	👍Fast-paced way to meet many customers.
13. User Groups *(similar to, but not the same as, Customer Councils)*	Groups of customers that meet on a regular basis to discuss technical issues. More frequently done over the Internet, as in list servers.	To address specific issues with a set of users. To identify trends in the industry. To hear the "buzz."	👍Relatively easy to arrange. 👎Sample may be skewed by those who enjoy joining user groups.
14. Professional Organizations	Membership in trade or professional organizations — for example, sponsoring an event or a research project.	To identify current trends in an industry or application area.	👍Easy to set up. 👎Requires commitment for the long run in order to develop trusting relationships.
15. Standards-Setting Organizations	Membership in professional or government organization responsible for setting industrial standards. Some are worldwide.	Understand the future of your industry. Able to influence the future. Network with early adopters and opinion makers. Meet industry gurus.	👍Discover future trends in your industry. 👍Good place to meet leaders in your industry. 👎Requires time and long-term commitment. 👎Your competitors are likely part of the group.
16. Lead Customers *(a.k.a. Lighthouse Customers)*	"Early adopters" is another term for these customers — not to be confused with technophiles. Identify customers who must have the latest technology, processes or products in order to meet their business needs.	To obtain input about the future of your industry. To identify emerging needs.	👍Good source of future needs. 👍A good way to anticipate the needs of late adopters. 👎One point does not make a trend — i.e., risk of being misled by a single point of reference. 👎The future market does not happen; it was a fluke.

Step 2: Organize Your Visits

Type of Visit	Description	Application	Pros & Cons
17. Cross-Functional Strategic Development	A formal agreement to jointly develop a product who will benefit the customer of the relationship, i.e. your company and that of the customer agree to develop the new product jointly.	Formal participation by the supplier in the development of a product for the customer. Part of the supply chain model, where suppliers work closely with customers.	👍Best position to address the needs of "leading customers." 👎Complex legal arrangements. 👎Requires considerable resources.
18. Industry "Opinion Makers" *(also Scientific Advisory Boards, such as health care)*	Systematically identify those individuals who influence the opinions of a market segment of a specific industry. For example, authors of popular industry books, keynote speakers in industry conferences, selected academics, etc.	Spot trends in the industry. Obtain specific information about an industry.	👍Good way to identify industry trends. 👍Understand those who shape the opinions of an industry or market segment. 👎In some cases you may need to pay honoraria or other fees.
19. "Service" Day	Develop a programmatic set of contacts using your service and repair organization. In the case of software, use the tech support organization. Tap customers when they bring or send equipment for warranty service.	To debrief customers who, by definition, are familiar with your product. To obtain specific information about your current products.	👍A convenient way to get customers, as they are already in contact with your company. 👍Real-time feedback on your current products. 👎Generally limited to your own products.
20. "Plugfest"	Similar to a user group, but intended to be a "hands-on" event. Customers bring their equipment (or you provide it) and start using it and exchanging information.	Ease-of-use exploration. Use of specific capabilities or features of a product. Discovering unmet or unarticulated needs. Opportunity to create long-term relationships.	👍A safe and perhaps fun environment. 👍No competitors; it's your agenda. 👎May be difficult to organize. Requires logistical support.
21. "Hire-A-Customer"	Some organizations hire a real customer — someone between jobs or outright from the marketplace. You can also hire them under a contract or even as consultants. Their role is simply to behave, represent and advocate for the customer within your organization.	To provide a constant reminder of the voice of the customer. To be available to provide immediate feedback to your team.	👍A fine way to bring the customer inside your organization. 👍Convenient and available. 👎Over time, they need to be replaced because they no longer feel like a customer.

Step 2: Organize Your Visits

Type of Visit	Description	Application	Pros & Cons
22. Customer Panels *Due to their importance for customer requirements, Customer Panels are fully described in Critical Tools.*	Meeting with a small group of customers on a specific topic or subject. Similar to Focus Groups, except that the customers interface directly with you and your team, not through a trained facilitator. Generally a single event, such as going to a city and inviting customers from the area to attend.	To obtain input from a representative group of customers. To listen to several customers in one place.	👍Good way to create focus. 👍Good way to create synergy among customers. 👍Excellent tool to enable technical personnel to interact with customers. 👎Takes much coordination. 👎May be expensive.
23. Focus Groups	A third party is in charge of the conversation. You are not in direct contact with your customers.		

Conclusion

There is no shortage of forums to meet and understand customers. With a little imagination, you can use the above list of ideas or you can create your own settings. All you need is your commitment to give customers the priority they deserve.

Step 2: Organize Your Visits

Alternative Sources

There are other sources of information about your customer, which can supplement, but not replace customer interviews. The following list provides a set of examples of customer contacts that do not take much time or money. Challenge your team to invent clever ways to stay in touch with your customers.

1. Subscribe to the top trade magazines that your customers read.
2. Help out in your customer service center (telemarketing centers, application center or call center).
3. Work with your service organization, and in the case of software, the installation or application organization.
4. Work with local schools and universities or establish a relationship with the leading educational institutions in your industry or the industry of your customers.
5. Join professional organizations in your area (chapters, sections, divisions).
6. Field feedback notes (field memos or the documentation that your distributors or sales force submits to the head office).
7. Volunteer to give "tours" to visitors or to assist customers when they come to your office.
8. Attend sales training sessions.
9. Make telephone calls to your customers.
10. Use bingo cards and other request forms from your customers.
11. Use warranty cards.
12. Connect through the Web, such a blogs, wikis or user groups.
13. Find internal users of your products, such as your suppliers.
14. Use networks.

Visiting Your Customer Need Not Be Expensive

You and your team can create various ways to obtain valuable customer input that cost very little, and in some cases nothing at all. You must be willing to be innovative, and allow your team to collaborate — be open to new ideas. The purpose is to obtain valuable input from your customer, and not having the budget to do it may be a small barrier, but it does not mean that you give up.

Visiting Your Customer with Minimal Effort: Example 1

Suppose that you have 10 members in your team (organization). Seven are from engineering, and the remaining three are from marketing.

Suppose also that you make a simple commitment: Each of you calls one customer per week for the next three months.

The simple math is that your team will make 10 calls per week, which means 40 calls per month, resulting in a total of 120 calls.

To be conservative, cut the number in half, although one call per week will not disrupt anyone's workload. So you have 60 calls — not to shabby!

Naturally, to make the above example work, you will need to organize and develop a plan as outlined in this book. Nevertheless, the information you can derive from 60 interviews is splendid, not to mention very useful.

Step 2: Organize Your Visits

Visiting Your Customer with Minimal Effort: Example 2

Suppose that the most important trade show in your industry is forthcoming. What if you form a team of three people, two from marketing and one from engineering, and go to the show.

But you also establish a purpose, for example, to visit the booths of all your competitors and obtain all possible information — legally and ethically, naturally.

Once you are back in your office, the three of you spend three hours debriefing. You will be pleasantly surprised at the quality and quantity of valuable information.

Another alternative would be to go to the trade show and ask people who come to your booth for a 15-minute interview in exchange for a small reward. During the duration of the trade show you might get 10 or more interviews — not bad for minimal effort!

Visiting Your Customer with Minimal Effort: Example 3

Work closely with your service, technical support or customer service organizations. Collaborate with them to obtain names and contact information from customers who are willing to be interviewed. Then, set about working with your marketing and engineering teams to call these customers and interview them. While telephone interviews are not the best way to interview customers, you nonetheless get valuable information. Consider this approach: Ask customers if they wish to be interviewed face-to-face as part of your product development efforts — they may be agreeable.

Conclusion

The above three examples were actually implemented by the authors; they are real examples. The point is that with minimum effort your organization can obtain very valuable information. It does not take a lot of money or a lot of time. Be open to new ideas, and be willing to innovate. Do not let the lack of budget keep you from talking to the most important person in your business.

Step 2: Organize Your Visits

Determining the Cross-Section

Using a cross-section ensures that you achieve a balance in your sample selection. It's how you ensure a diversity of sample within a set of customer interviews.

Use a cross-section when choosing whom to interview to ensure that your sample truly represents the targeted customer set. For example, you wish to talk to engineering managers, yet there are engineering managers in small companies as well as large ones. Likewise, there are engineering managers in Europe and in Asia, and so on. Thus, the cross-section helps you refine the mix to get a balanced set of input. It also helps you define who you are *not* going to interview, so that your sample is not biased in the wrong direction.

The cross-section will map your entire sample so you can evaluate its depth and breadth. Remember that by definition your sample will be small, between three and 24. So make certain that these individuals are a true representation of the specific customer set you are targeting.

Example: Mapping Visits for Relevant Cross-Section

Group Dimension and Customer Archetype

Group	Customers					
Sample Selection Sub-Groups	Current satisfied customers	Current non-satisfied customers	Lost or competitor's customers	Customers gained from competitors	Not yet customers	Total
Small companies (<$100M per year)	1		1			2
Medium-size companies (>$101M, <$1B)		2	1	1	2	6
Functional types (title of customer to be interviewed)	3	2	5	2	2	14
Total Interviews	4	4	7	3	4	22

One approach to ensure that you have the right cross-section in your sample selection is to focus on company size and then on the customer archetype. For flexibility, you can use other group dimensions to define your sample selection, for example:

- **Industry segment, such as semiconductor or telecommunications.**
- **Geographic segments, such as Western States or Europe and Asia.**
- **Frequency of use of your product, such as frequent users, occasional users and rare users.**
- **Technology maturity, such as lead customers, early adopters, laggers, etc.**
- **Application, such as research or design, manufacturing.**

The number of interviews you will conduct is driven by your objectives. In this example, the team will conduct 22 interviews.

Step 2: Organize Your Visits

Example: Cross-Check Visits for Relevant Cross-Section

Target Company and Function

Visit ID#	Target Company	Target Function (Who)	Venue (Where)	Number of Interviews	Cross-Section Criteria
Used for tracking purposes	Name or type of company	Name and title	Where the interview will occur	Interviews with this profile	How the selected criteria is met
Small #1	Acme Electronics, small manufacturing company in Asia	Sue Wong, Engineering Manager, Electrical	At the Asia-Comm conference in Singapore	1	Interview two engineering managers in small contract manufacturing companies in Asia
Small #2	Ace Electronics	Fred Trang Engineering Manager, Mechanical	At Ace Electronics in Shanghai	1	See above
Large #1	Valley Electronics	Joe Hallett VP of Engineering	At Valley Electronics in Carlsbad, Cal.	1	Interview two senior engineering managers in large manufacturing and design companies
Large #2	Rapidtech	Mary Johnson Director of Engineering	At the IEEE Power Division meeting in San Diego, Cal.	1	See above

The above table demonstrates how to ensure that your sample selection meets the right cross-section criteria you developed for your set of interviews. This cross-check is the last step in your sample selection, and it's intended to map all your planned visits against the criteria you and your team developed for your project.

In this example, the focus is on Company Name and Function. The Visit ID# not only allows you to track all your interviews, but also you can view them at a glance to ensure clarity and accuracy. The venue, where the interview will take place, also plays an important role. Cross-Section Criteria represents the how a single interview contributes to the overall fulfillment of the criteria you have selected — every interview in this round should add up to you meeting your criteria.

For flexibility, you can redefine the headings or create your own. The important thing is to ensure that your sample selection is balanced and representative of the types of customers you wish to understand.

Customer Contact Planner Tools

Templates for Organizing Your Customer Visits

Introduction

Attention to the logistics is critically important when scheduling a series of visits. In the following pages you will find a set of templates, which you may use to plan and schedule your trip.

The templates in the following pages can be filled out, used as checklists or to help you develop your own template to best suit your needs.

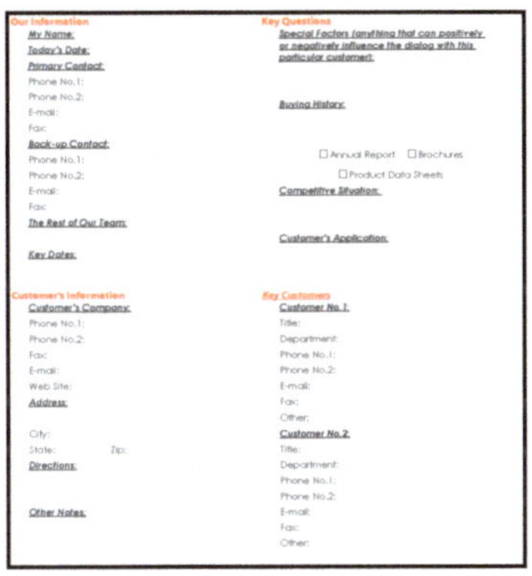

1. Contact Information Checklist Template

1. CONTACT INFORMATION CHECKLIST TEMPLATE:
Use this form to consolidate all relevant information about a particular customer contact. It helps you obtain critical information about your customer so you can be better informed during the interview. Use only the sections of the form that meet your needs; it is not necessary to fill out the entire form, but we recommend that you address every item in the form. Complete one of these forms for every customer. You and your interview team should fill out the information together to ensure clarity.

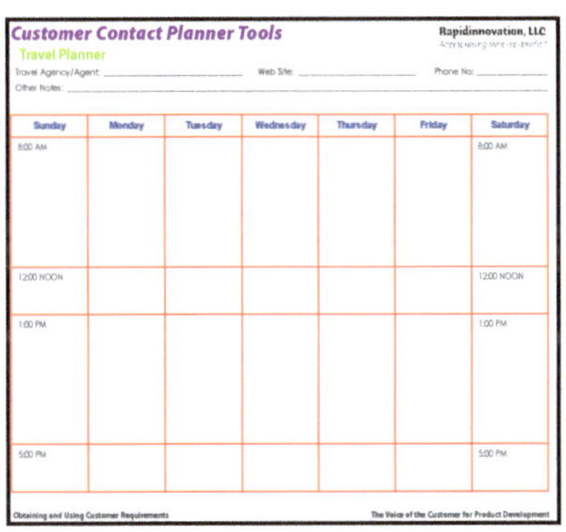

2. Travel Planner Template

2. TRAVEL PLANNER TEMPLATE:
Many times you schedule a visit without allowing enough time for travel from the previous visit. This tool will help you schedule a full week of visits. It is a practical way to map your contacts and check for omissions, overlaps and other scheduling problems.

This is a good opportunity to test your logistics; for example, travel time or potential delays.

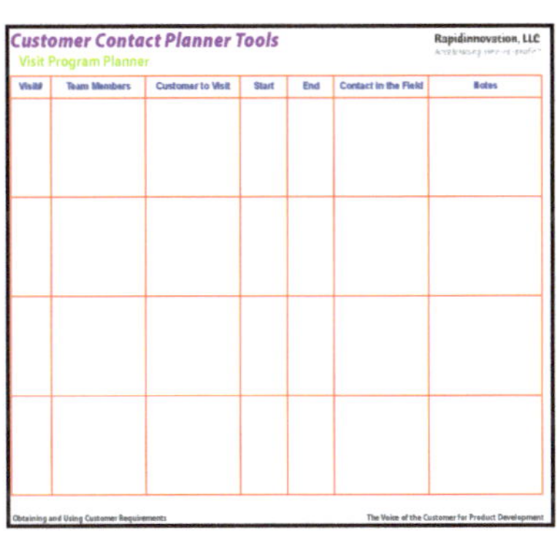

3. Visit Program Planner Template

3. VISIT PROGRAM PLANNER TEMPLATE:
This template will help you map all the contacts for a set of visits. Often there are several teams from your company going to the field. This represents a logistical challenge; for example, which team goes to which interview. This form will help you keep track of the number of teams and customers so you can view it on a single page. Be certain to develop this form in full collaboration with all the interview teams such that everyone is fully informed and aware of their responsibilities.

Step 2: Organize Your Visits

Contact Information Checklist

Our Information

MY NAME: ______________________

INTERNAL CONTACT #1: ______________________

Phone #1: ______________________

Phone #2: ______________________

E-mail: ______________________

INTERNAL CONTACT #2: ______________________

Phone #1: ______________________

Phone #2: ______________________

E-mail: ______________________

THE REST OF OUR TEAM: ______________________

____

____

KEY DATES: ______________________

____

Special Factors

POSSIBLE POSITIVE OR NEGATIVE INFLUENCES ON THE DIALOG WITH THIS CUSTOMER: ______________________

____

____

BUYING HISTORY: ______________________

____

____

❑Their Annual Report ❑Brochures
❑Product Data Sheets

COMPETITIVE SITUATION: ______________________

____

____

CUSTOMER'S APPLICATION: ______________________

____

____

Customer's Information

COMPANY: ______________________

Phone #1: ______________________

Phone #2: ______________________

E-mail: ______________________

Web Site: ______________________

ADDRESS: ______________________

City: ______________________

State: __________ ZIP: __________

DIRECTIONS: ______________________

____

____

OTHER NOTES: ______________________

____

____

Key Contacts at the Customer's Site

CUSTOMER #1: ______________________

Title: ______________________

Department: ______________________

Phone #1: ______________________

Phone #2: ______________________

E-mail: ______________________

Other: ______________________

CUSTOMER #2: ______________________

Title: ______________________

Department: ______________________

Phone #1: ______________________

Phone #2: ______________________

E-mail: ______________________

Other: ______________________

Step 2: Organize Your Visits

Contact Information Checklist (cont'd)

TODAY'S DATE: ___________

Pre-Visit Checklist

PREPARATIONS REQUIRED: ___________

OUR PROMOTIONAL MATERIAL: ___________

- ❑ Audio-Visual Presentation
- ❑ Demo Equipment, software, etc.

OTHER DETAILS: ___________

Discussion Guide

BASIS FOR STARTING THE CONVERSATION:

QUESTIONS OR THEMES TO BE DISCUSSED:

#1: ___________

#2: ___________

#3: ___________

Objectives of the Visit

PRIMARY GOALS:

#1: ___________

#2: ___________

#3: ___________

Anticipated Commitments

OURS: ___________

THEIRS: ___________

Step 2: Organize Your Visits

Example: Travel Planner

Travel Agency/Agent: ____________________ Web Site: __________________ Phone: ______________

Other Notes: ____

____

SUN	MON	TUE	WED	THU	FRI	SAT
8:00 AM	Breakfast with salesperson, in the cafe at the "Acme Hotel" by the airport 45 minutes travel to customer site 9:45 arrive a customer site 10:00 start meeting with customer	8:00 pick up by the salesperson at the front door of the "Acme Hotel" 45 minutes travel to customer site 9:00 start setting up demo equipment 10:00 start demo				8:00 AM
12:00 NOON	Lunch with customer (salesperson to make reservations					12:00 NOON
1:00 PM	1:00, 50-mininute travel to next meeting 1:50 arrive at customer site 2:00 start meeting with customer					1:00 PM
5:00 PM						5:00 PM

Important

Notice the attention to clarity. Be specific.

Any misunderstanding will result in aggravation, which will interfere with your interaction with your customers.

Do not leave anything to chance when developing a schedule for your customer interviews.

Step 2: Organize Your Visits

Example: Visit Program Planner

Template

Visit #	Team Members	Customer to Visit	Start	End	Contact On-Site	Notes
1	Roy P. Mary A.	Acme Electronic	May 15	May 16	Fred A., Salesperson Roger B., Engineering Manager with Acme Electronic	This visit will include a demo to explore the ease of use challenges (see discussion guide)
2	Roxanne F. Peter G.	American Electronics	May 15	May 16	Peter F., District Sales Manager Mary R., VP of Engineering at American Electronics	Meeting may include several engineers
3	Roy P. Mary A.	Technology Associates	May 19	May 20	Al B., Salesperson Mike F., Engineering Lead	Use same demo equipment as in visit #1. Needs to be updated before the visit.

Step 2: Organize Your Visits

Discussion Guide

Topics and Questions for Customer Contacts

A discussion guide is an essential tool to ensure maximum productivity from your customer contacts. It is the script, or roadmap, for the set of topics and questions you want to discuss with your customers. It serves as a tool to facilitate a smooth flow of information from your customer to you. We recommend that no customer visit be done without a discussion guide.

Benefits of a Discussion Guide

- It ensures that the time you spend with your customers is used for maximum efficiency and that all the important topics are covered and not missed.
- It keeps you and your team members from straying into useless discussion.
- It provides a natural flow — for example, from the abstract to the tactical.
- It ensures consistency across all meetings with different customers. It ensures that all the questions are asked in the same sequence and priority.
- It ensures that all interviewers or teams ask the same questions and in the same priority — this is helpful when teams travel abroad to make interviews.
- It provides an initial framework for processing the information when all the interviews are concluded. Imagine 50 pages of notes in a random sequence. The discussion guide provides a starting point for consolidating the information.

General Rules for Discussion Guides

We recommend that your discussion guide follow a sequence that starts at the abstract level and ends in specific areas. For example, start by asking "How is business?" and end with "What are the current levels of sales?" This avoids pushing the customer into a corner.

1. "Why" is always more important than "what." The critical issue is to understand why customers behave as they do. Asking "What do you measure?" is not as valuable as asking "Why do you measure?" The first question will result in a short, cryptic response, while the second will yield valuable information such as, "We measure to ensure quality in order to avoid defects." *Consider this example:*
   - Customers do not really want a "stapler." They want to organize information contained on paper. In this example, "the what" will yield you a better stapler. The "why" will yield more opportunities — i.e., how to help customers organize information contained on paper.
2. Avoid the interrogation mode, when you start asking questions in a fast sequence, machine-gun style. This annoys customers, and does not allow for the proper dialog to take place.
3. We suggest that you dry-run your discussion guide (some call it a pilot test). For example, one of your team members can play the role of a customer. This will validate that the flow is correct and that the questions are properly phrased. Further, you can test it with an actual customer but discard the results. The objective is to test the validity of the guide, not to collect information.
4. Keep it simple and short by prioritizing the topics to no more than three per each hour of visit. When it comes to talking to customers, everyone in your organization wants their questions answered. Unfortunately, this will result in an unmanageable interview, too many topics and too many questions. Rather, we suggest that you use clearly defined criteria to prioritize the topics of conversation.

Essential

"The discussion guide provides the sequence, rhythm and consistency to ensure successful interviews."

5. The discussion guide is not a straitjacket; it is simply a direction that allows for a smooth-flowing dialog with the customer. You must use your judgment to decide when a topic is exhausted, when the customer cannot give you additional information and even when you must deviate from the guide.

A Few Questions about Questions

The following questions might help you refine the types of questions you formulate. The challenge is always the way your customer will interpret your question — i.e., the connotation that she will assign to your question based on her environment, experience, mood, etc. Even with the best of intentions, we face the risk of asking the question but obtaining the wrong answer.

1. Is there bias in the way you have formulated the question? Said another way, does your question predispose a specific answer?
2. Can the customer answer the question? For example, avoid asking a technical question to a non-technical person.
3. Do the words in your question have a single meaning, or are they too easy to interpret in different ways?
4. Are your questions arranged in a logical fashion that makes sense to the customer? While they may seem logical to you, they may be confusing to your customer.
5. Is the question really needed? Will the answer help you achieve the objectives of the interview?
6. Are you using jargon or a language that the customer may not understand, such as acronyms that are only used in your company?
7. Is the question formulated in such a way as to generate the right information?
8. Will your question become more effective if you use a visual aid such as a diagram or picture?
9. Are there assumptions that are not mentioned but should be?

A Few Topics that Work Some of the Time

These are questions or topics that have worked for many users. They may provide a starting point for you:

1. Identify tasks that the product must meet, and why.
2. Describe how the product will be used, and why.
3. Identify unsolved problems.
4. Identify "new" problems.
5. Identify general likes.
6. Identify general dislikes.
7. Generate a list of trade-offs.
8. Visualize the product in use.
9. Identify unmet or unstated needs.

Step 2: Organize Your Visits

A Few Approaches that Work Some of the Time

These are questions that you might consider depending on the nature of the topic you have selected. This table is for illustration purposes and is intended to help you create your discussion guide, not replace it.

Objective	Sample Questions
1. To understand the customer's current situation	• Tell me about your growth plans? • Please describe your current manufacturing process? • Tell me about your current new product development process? • What is your current competitive situation? • What new products are you developing?
2. To identify future needs	• Please describe the ideal system you would use in the future? • What critical improvements would make to your manufacturing process to meet your future needs? • Describe the ideal system, what would be the "perfect system" for you? • What critical changes do you anticipate in the next three to five years? • How will your industry be different five years from now? • How will your company change in the next five years?
3. To identify current problems	• What worries you most about your manufacturing process? • What is the biggest problem you face in your system? • What impact will the new regulations have on your operation? • What keeps you awake at night? • What worries you most? • What is your number one problem?
4. To address operational needs	• What are your top three objectives for this year? • What critical obstacles do you face this year? • What keeps you from optimizing performance this year? • What barriers must you overcome this year
5. To help the customer prioritize	• What single improvement would you make to improve efficiency? • What one solution would help you achieve your business objectives this year? • What is the biggest challenge you must solve in the next three months? • What is the one area of your operation that creates the most problems for you?
6. To help the customer visualize innovation.	• What if you could double your productivity? • Imagine what the ideal solution to this problem would be? • What would be the perfect solution to this problem • What if you could eliminate the bottleneck in your operation? • What would be some possible solutions to this problem

Step 2: Organize Your Visits

Level 3 Detailed Process: Developing a Discussion Guide

1

Set the Foundation

Ensure that the objectives for the customer visits are documented and final.

Circulate the objectives and any other relevant material to the members of your team and any other relevant stakeholder.

2

Call a Brainstorming Meeting

Decide who needs to attend based on the objectives and each person's ability to contribute.

Schedule the meeting for at least three hours. This may reduce attendance, but quality is more important than quantity.

Secure a conference room with ample space.

Important

Define and prioritize topics for the meeting conversation. A good way to start the brainstorm is to ask your team members to write down all the questions they wish to ask their customers (use yellow stickies).

Time Frame

Brainstorming Meeting

3-4 Hours

3

Start Brainstorming Meeting

Start by reviewing the "purpose" of the interviews

Next review the "objectives" for the interviews.

Allow for some discussion to ensure clarity.

Set the stage by asking everyone to write down on a yellow sticky (Post-It Note) question they would wish to ask customers.

Ask everyone to read their yellow stickies and encourage the rest to write additional questions.

4

Brainstorm

Use the Affinity Diagram method or any other formal framework for brainstorming.

- Brainstorm all possible topics.
- Organize the topics or questions.
- Consolidate the topics.
- Prioritize the topics.

You should end up with no more than three to five topics. Prioritize based on importance to the objectives and using a clear criteria for selection.

Important

Generally, you will need one hour to cover three topics, or perhaps fewer, during a customer interview. Remember that you need to allow the customer to talk, and this takes time. It is not a survey, it is a conversation where you need to probe and let the customer talk and think.

See Also

Affinity Diagrams

Continued

Step 2: Organize Your Visits

Cont'd

5

Prioritize

At this point you should have a number of yellow stickies, perhaps 100 or even more. You will also have a combination of questions, topics and comments.

You now need to prioritize. Your objective is to end with three to five topics for general areas of discussion.

See Also Critical Tools

See Also Hierarchy of Topics and Questions

6

Develop the Questions for the Topics

Now that you have the topics, you can proceed to develop the questions for each topic.

Always start at the strategic or abstract level, then work your way down to the tactical areas.

You should also have follow-up questions. This will help you pursue the topic to any depth that you need.

7

Create Your Discussion Guide

A discussion guide is really a script that will guide you through the entire interview with your customers. It's not a set of questions, but rather a framework for dialog.

The discussion guide always starts with the introduction, where you explain to the customer the purpose of your interview, establish the context and address any issues or concerns. Write out the introduction to help you remember and also to ensure that all your interview teams use the same format.

Next, address each of your topics or themes contained in your discussion guide.

Conclude the interview by thanking your customer, answer any concerns and reviewing any action items.

8

Test Your Discussion Guide

Use one of your team members; ask her to play the role of the customer and test the guide.

Schedule a meeting with an actual customer, but for the purpose of testing your guide.

Important!

To keep the interview focussed, develop a set of questions that matches the topics you've prioritized.

To allow for a more dynamic conversation with your customers, you could use forgo scripted questions.

Note, however, that we have found that if you are not yet experienced in customer interviews you ought to use a set of questions.

9

Finalize Your Discussion Guide

Create the final version of your discussion guide, distribute it to all team members, and ensure that everyone understands how to use it.

Continued

Step 2: Organize Your Visits

CONT'D

10

USE THE GUIDE IN YOUR INTERVIEWS

Follow the guide, but use your judgment. Notice what works and does not work with your guide — for example:

Does the flow of questions make sense to the customer?

Are your questions understandable? Or do you need to spend time explaining the question?

Does the question yield meaningful information? Or do you end up with trivial responses?

11

DEBRIEF AND IMPROVE

After each interview, talk to your team members about ways to improve your guide.

Ensure that all guides are changed. For example if there are three teams doing interviews, all the guides must be the same.

12

CONTINUE YOUR INTERVIEWS

Step 2: Organize Your Visits

Example: Hierarchy of Topics and Questions

Topics are simply high-level themes you wish to present to your customers. Themes are not questions; you formulate questions as you interact with your customers. We highly recommend that at least the First-Level Topic be phrased as a theme. Some teams choose to write actual questions related to the topic in the second and third levels. This approach is acceptable, but there is a risk that your interview teams might use the questions as a survey, which this form is not. You and your team must become comfortable with formulating questions during the interviews.

First-Level Topics (Themes)	Second-Level Topics (Themes)	Third-Level Topics (Themes)

Example #1

First-Level Topics (Themes)	Second-Level Topics (Themes)	Third-Level Topics (Themes)
Reaction to the new menu scheme.	Probe for any obstacles when moving from the icon to the point-and-click.	Probe to identify any preferences.

The above example shows a simplified discussion guide (one topic only). Note that in this example there are no actual questions; rather, it focuses on the specific topics for discussion, and what specific areas to probe. This is acceptable if you have some experience in interviewing customers; otherwise, we suggest that the Second Level Topics be questions.

Example #2

First-Level Topics (Themes)	Second-Level Topics (Themes)	Third-Level Topics (Themes)
Identify the purchasing process, and the role of the end user in the process.	Ask him to remember the last time he purchased a "left-handed widget," and describe any hesitations and reservations he had. Ask about any risks he had to take when he last purchased a "left-handed widget."	Describe the forms you need to fill out to be able to purchase a "left-handed widget."

In this example the general topic outlines the intent of the interview; i.e., the purchasing process. The Second Level suggests the specific questions, but does not phrase the questions. It is left to the interviewer to formulate them. In the Third Level there is an actual question, perhaps to ensure that it is covered, or to obtain specific information.

Example #3

First-Level Topics (Themes)	Second-Level Topics (Themes)	Third-Level Topics (Themes)
Product Alpha, usage experiences.	As a user of product Alpha, what do you like best about it? What do you like least? Please describe your experiences when using product Alpha.	Probe for the level of satisfaction with customer service for product Alpha. Ask him to describe his experiences in using the Web page for product Alpha.

Step 3: Interview Your Customers

Focus on the Customer

Face to Face with Your Customer

This is the moment where all the planning, traveling, and expense pay off: when you are doing the actual interview with your customer. You are now familiar with the discussion guide, and are comfortable with the topics for discussion. You have clearly defined roles with your partner; for example, who will make the introductions, who will explain the reasons for the visits, who will take notes, etc.

You were also very careful about the logistics. For example, you allowed plenty of time to travel to the customer's site, so as to prevent being late due to traffic congestion or some other contingency.

Speaking of logistics, remember that thing can, and will go wrong. You and your partner need to react and correct so as to salvage your interview. When something goes wrong, it's no time to talk about whose fault it was. Rather, set about fixing things and interviewing your customer.

Communication Skills

You are ready to focus on your customer and practice all the rules of effective communication. For example, listen with a purpose, never "debate" the customer, and expect the customer to do 90 percent of the talking.

Additionally, advanced communication skills, shown later in this guidebook, will greatly increase your ability to hear your customer's message.

Facilitation Skills

Your role is to enable the customer to talk, to express his needs using his language. The interview, then, is not a conversation, nor is it a dialog. It is a facilitated interview, where you enable the customer to articulate his feelings, needs, requirements and any other valuable input.

You must create a "safe environment." Ensure that the customer does not feel threatened or ill-at-ease. Reassure that the information will be held in confidence. Also that his supervisors or managers are in support of the interview. Your demeanor should also be easygoing and relaxed. If you are tense, your customer will be, too.

The interview is all about getting the customer to talk about himself, his world, his needs, his problems. Frankly, the interview is not about your product or your company.

Note-Taking

The process favored in this guidebook relies 100 percent on your ability to take notes. You must be comfortable note-taking and proficient in the art and science of note-taking. We cannot stress this enough!

When taking notes, write down everything of relevance. Even if it seems like something you will remember later, you likely will not.

Essential

"The process favored in this guidebook relies 100 percent on your ability to take notes."

Step 3: Interview Your Customers

Customer quotes are essential, so be sure to capture them verbatim. Also, mark them in your notes so you can separate your thoughts from the customer quotes.

> "I wish someone would make it easier to switch hard disks in my desktop computer."

As shown above, a quote is captured verbatim, and it is flagged as such — notice the quotation marks.

A Customer Quote is a Customer Quote

It is absolutely critical that you identify those passages that are customer quotes; that is, when you are taking notes and you capture a quote from your customer, identify it as such. Customer quotes are indeed the voice of the customer, and should not be confused with your notes. This is an essential rule for this methodology. By the same premise, do not mark your notes as a customer quote.

Methods of Capturing Customer Quotes

> **Essential** "When doing audio, video or still photos, be certain to ask permission first."

Audio Taping is Critical

When interviewing a customer, you are very busy observing, asking questions and taking notes; you cannot possibly capture all the nuances and important details. Plus, you will be doing more than one interview, traveling from place to place; consequently, you are likely to forget much of what was said, particularly the critical information.

After many years observing teams try to capture information during an interview, the authors have concluded that audio recording the interview is the only way to obtain maximum value. We are always amazed at the additional information you obtain when you review the recordings, very valuable information is contained in the recording, information that you were not able to capture in your written notes.

We highly recommend that you equip yourself with small, digital recorders (be certain that it is a digital recorder); they are not expensive and are very convenient due to their small size. Be certain to equip all the Interview Teams with recorders as well, it's a wise investment and you can use them in future research as well.

Inform the customer that you wish to audio tape because their input is very valuable to you. In most cases this will not be a problem. Most customers understand your need to record the information. In very rare occasions, the customer will not allow you to record the interview, in such occasional cases proceed with the interview and take copious written notes.

Having a recorder will not interfere with quality of the information, customers will soon forget that they are being recorded and express themselves freely.

Transcriptions from Audio Tapes

Having the audio will help you later when you are preparing to process the information. An additional level of knowledge is obtained when you have all the recorded interviews transcribed so you and your team can read them for analysis. The ability to read, highlight and review typed information is very helpful to gain maximum insight into the voice of your customers.

You can have the tapes transcribed by companies who specialize in this type of work. Search for "Transcription Services" in your local area or on the Web. These companies serve the legal industry such as attorneys, lawyers and other legal entities. One advantage of digital recorders is that you can e-mail the audio to the transcription service. Then, it does not make any difference where they are located.

We highly recommend that you have your audio recordings transcribed so you and your team can read the interviews. This is the best-case scenario. It is worth the effort and any additional time you spend.

Step 3: Interview Your Customers

Videotaping

Videotaping is another excellent way to capture information. The added benefit of the facial expressions, the environment and other subtleties make it a good investment. As with audio, ensure that it is acceptable to your customers.

One additional benefit of video is the ability to share it with your peers back at the factory. This has powerful impact, and it must not be underestimated. Plus, you can show segments when making your product proposal.

Still photos

Still photos are also a good way, albeit limited. They do provide a good record of your customer's face, surroundings, and other installations. With digital cameras it is now much more convenient. The same rule applies: ensure approval from your customers.

Barriers

Occasionally you will find customers that object to audio, video or still photographs. We recognize that this does happen on occasion. Never the less, we encourage you to ask, you would be surprised how many times customers agree. When you ask, reassure your customer that you respect their information and also that you will use it only for the purposes of developing your new product.

Suggestions to overcome resistance:

- Some times it is best to meet the customer away from their office, which makes audio or video more likely.
- Offer to sign a Non Disclosure Agreement (NDA) to protect the confidentiality of the information
- Reassure the customer that the information will only be used by your team, also that you will not — in any way — quote or use names, brands or intellectual information outside of your team.
- Offer the customers a camera or a video camera so they can document the information, for example a use case or the layout of their lab or factory floor. This reassures your customer that they can control what is documented

Finally, there will be a few times where the customer simply cannot allow the electronic documentation due to company policy or other restrictions. Simply accept the constraint and proceed with the interview.

Step 3: Interview Your Customers

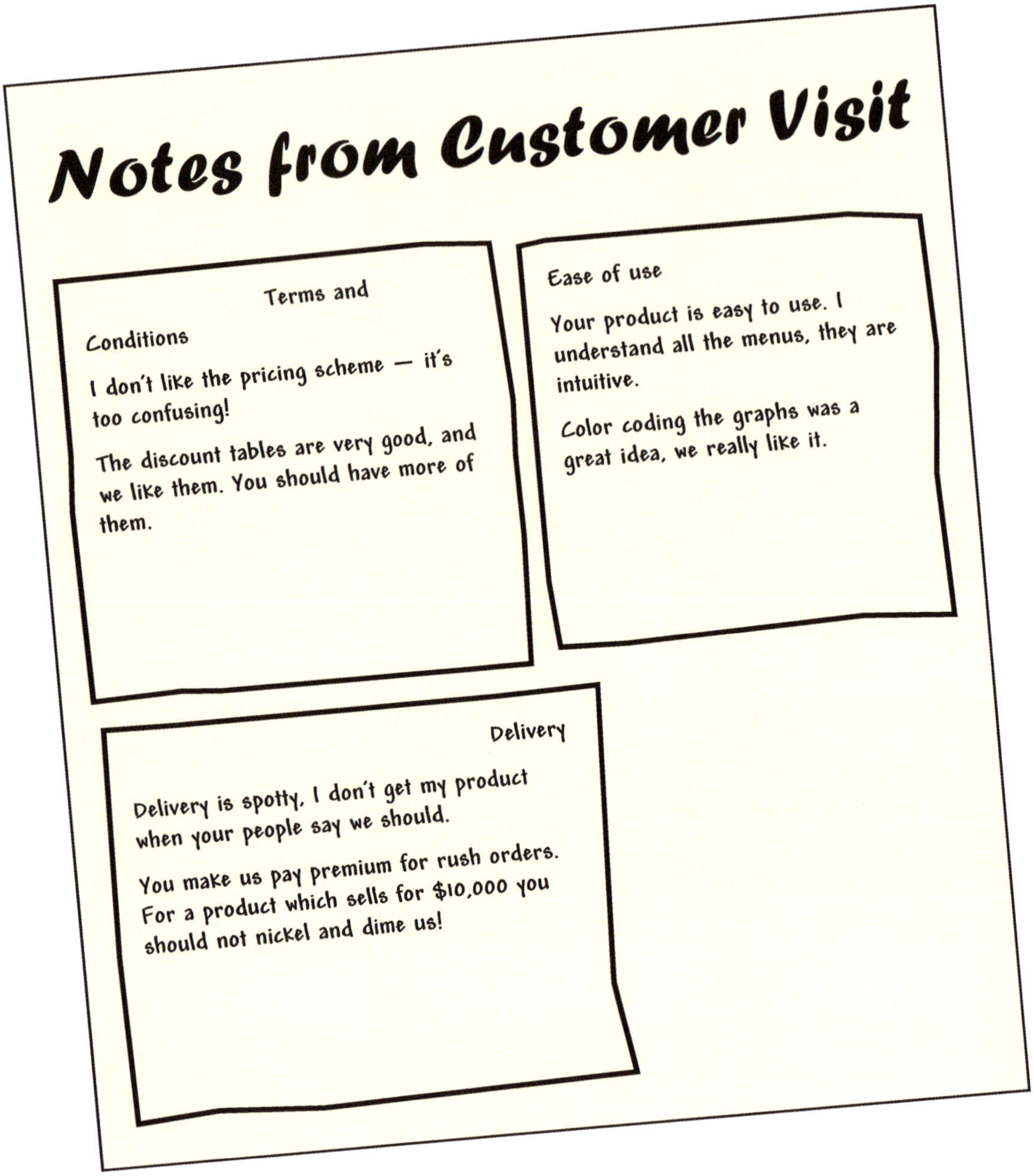

Note-Taking Tip

Humans in general do not talk "linearly"; that is, we do not start a topic and continue until finished. Rather, we jump from topic to topic. This cannot be changed, so you should learn to live with it.

So, how do you take meaningful notes when your customer jumps from topic to topic? One technique is to draw several squares on your note paper, then use each square for a different topic, as shown in the example above. You will find that you will be able to jump from square to square and keep up.

Step 3: Interview Your Customers

Basic Communication Skills

Communication

Not surprisingly, your ability to communicate is essential to understanding your customer. This calls for a conscious effort on your part to refine your communication skills, simply agreeing that communication is important will not suffice. You need to learn and practice good communication.

We have divided this section into two parts. First is the basic communication skills needed for your interviews. Second, we present advanced concepts in communication.

Observing

Much can be learned from customers by simply observing their environment. Start with the lobby, the conference rooms, and the overall ambiance of their office or factory. You need to concentrate from the moment you enter the building, and pay close attention to everything you see. Also, take notes of important observations; for example, is it an austere environment or perhaps luxurious, this tells you about the culture of your customer's company.

Observation extends to the realm of nonverbal clues. For example, note the attitude, style and surroundings. Do you detect a "dress code," be it formal or informal? Are your customers on a first-name basis, or are they more formal. Is there an obvious "pecking order," which might indicate a strong hierarchical arrangement?

Observation leads to insights into your customer's value system, which helps identify needs.

Active Listening

You must listen with intent; simply listening will not be enough. You must focus, and open your ears to everything being articulated. In short, listen with a purpose. Do not allow your biases and preconceived ideas to cloud your ability to listen. And above all, let your customer talk! As we mentioned before, you interview is neither a conversation nor a dialog. You are there to facilitate your customer's ability to articulate her requirements.

Verifying

Often, what we think we heard is not what was said — much less what was meant. This is why verifying is a critical skill for understanding your customers.

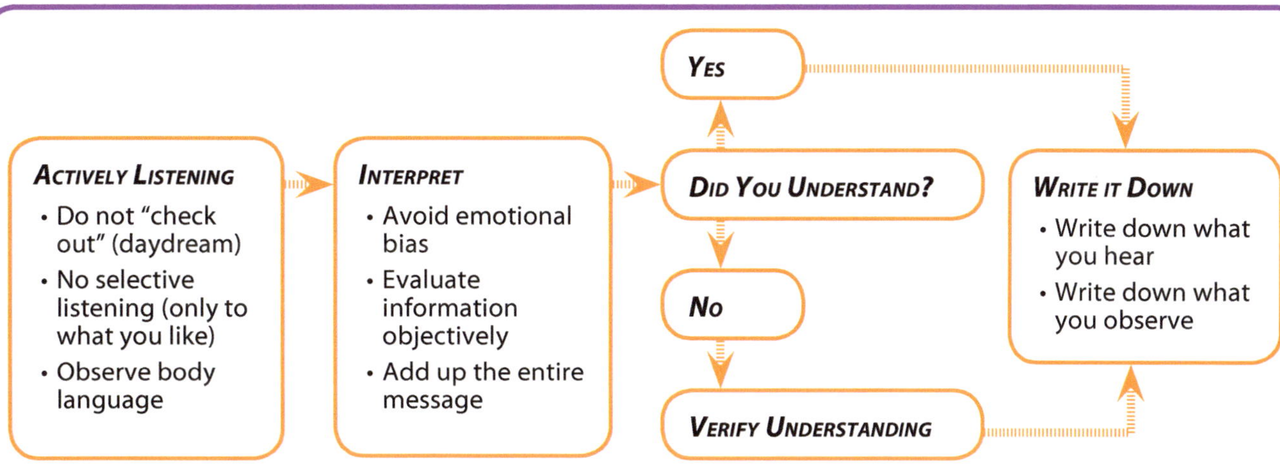

Active Listening is a Process

Step 3: Interview Your Customers

Verifying is simply your ability to listen and ask questions if you feel the message was not clear or you did not understand. Get into the habit of verifying; it is needed more often than we care to admit, and the voice of your customer is important. Thus, it is worth validating before you are no longer in front of your customer.

The technique for verifying is quite simple. Listen, then pause to analyze. Then summarize for your customer and seek confirmation.

Example of Verifying

- Your customer: "I don't like cell phones with little keypads."
- You verify: "Do you mean that you do not like small keys on your cell phone?"
- Your customer: "No, what I meant is that I don't like keys that can't be seen at night"
- You: "Oh! Now I understand."

Obstacles and Complaints

An obstacle is anything that impedes forward progress in your interview. For example, all of a sudden your customer shows impatience, or short temper. Obstacles will happen, and you need to be ready to deal with them so you can continue your interview. Obstacles can be divided into at least four categories. First is disinterest: Your customer simply is not engaged in the conversation. Her body language tells you that she is not happy being part of the conversation. Second is inability: When you are asking a customer a question she is not equipped to answer. For example, you ask a technician a question that should be asked of an engineer. Third is an objection: This is simply a disagreement; your customer has strong convictions or believes that what you said is not correct. Fourth is a complaint or dissatisfaction: For example, your customer believes that your products are not reliable. Any of the above categories may bring your interview to an unpleasant stop.

What to Do

Keep in mind that the "objection" may be the way your customer seeks additional information, or perhaps she is just thinking out loud. Finally, she may just be seeking reassurance. Therefore, an objection may not be what it appears, but simply the communication style of your customer.

The Move-On Step

The way to handle objections, obstacles and complaints is relatively simple.

1. First, keep your cool, stay calm and do not overreact.
2. Second, pause to analyze. This normally is called "dead air." Simply take a few seconds to think, take stock and avoid overreaction.
3. Third, verify to ensure that indeed it is a complaint or an obstacle. As we mentioned before, what seems to you as an obstacle may be your customer's way to obtain additional information.
4. Fourth, handle the issue accordingly. In the case of an obstacle, you need to use your judgment and handle it accordingly.
5. If it is a complaint, write it down so your customer recognizes that you give it importance. Then, reassure your customer that you will personally resolve the issue. Do remember that it is not settled until the customer says it is. Go out of your way to ensure satisfaction! What if the complain is about a product from a different division of your company. Matters not! When in front of your customer, you are the company — all of it.

A word about complaints. They are a source of new ideas, perhaps an unmet need. Also, complaints are an opportunity to improve customer satisfaction. And, finally, customers want to be heard: Let the customer talk!

Step 3: Interview Your Customers

Questions to Use, and How and When to Use Them

Introduction

Questioning is the most important tool to understanding customer need, and it serves as the primary vehicle of communication between you and your customer. Proper questioning enables the customer to provide you with the critical information, insights and emotions that will enable your team to provide a successful solution.

Asking the customer "What do you need?" or "What are your needs?" will not work. All of us have difficulty articulating our needs; furthermore, there are needs that are hidden. Some call them unarticulated or latent needs. It is best to focus on the "why." That is, why do customers do what they do? Asking why will give you the background you need to define the value that customers are requesting. The common example of the stapler comes to mind. If you ask "What do you need?" the answer will be a stapler. But if you ask why, the answer will be, "I need a convenient, easy-to-use way to organize information," which provides many more opportunities for added value.

Important: Questioning is the most important skill to understand customer needs.

Key Points about Questions

Everything must start with your objectives — i.e., what is it that you wish to obtain from your customers? Thus, the need for a discussion guide becomes obvious, as it enables you to carefully choreograph the type, sequence and emphasis of questions.

Remember that most of our communication has little to do with the actual words we use; rather it is the tone, body language, situation and other emotional variables. So we suggest that you avoid the "computer voice" when asking your questions. Instead, be spontaneous, sincere and, most of all, genuinely interested in your customer, who, after all, is as human as the rest of us.

Insights about Questions

1. A customer visit is not an interrogation, so you want to have a smooth flow of information with the customer.
2. It is not a dialog. If you think about it, your role in a customer interview is to serve as a facilitator, enabling the customer to articulate information. Approach it on the assumption that the customer will do 90 percent of the talking and you will do 10 percent, simply to enable your customer to articulate the information you need.
3. The customer interview is not the place to agree or disagree with the customer; leave that for another day. Instead, focus on understanding why the customer feels so strongly about the topic or deciphering how she came to feel the way she does.
4. One characteristic of humans is that we articulate information in a nonlinear fashion. For example, your customer will change topics while answering one of your questions. This is normal and human, and you should be prepared to facilitate by acknowledging the information and enabling the customer to get back on track. Note that in some cases, the tangents that crop up during an interview can provide valuable information, so be flexible and exercise judgment.
5. The follow-up is the secret. Your ability to think on your feet and follow threads during the customer interview will yield the critical information that does not surface during the first answer. There is a technique known as "peeling the onion" or "pulling the thread." This simply means your ability to continue asking questions as long as the customer is providing valuable information. The endpoint of a thread is when you get the root message from the customer, which is the information that will enable you to fully understand her situation. Therefore, you peel the onion to get to the root message.

See Also: The 5 Why's

Step 3: Interview Your Customers

6. Relate the customer to her situation, not yours. Your goal is to learn how to see the world through the eyes of your customer; it does not make sense to force-fit the customer into your point of view. So the use of phrases like, "Visualize your last experience with this product," "Imagine that you were using this produce" or "Picture you being able to solve this problem" are good preambles or set-ups to questions. Otherwise, the customer will not have a context or a point of reference.
7. Dead air is good — use it to your advantage. In every conversation, there are moments of silence that are generally short but feel like an eternity. In a customer interview situation, dead air allows your customer to think and organize her thoughts. It also forces customers to dig deep into their experiences. Do not rush the customer; allow the dead air to do its job. Don't be in a hurry or panic because there is silence; rather, get ready to take notes.

Open-Ended Questions (Open Probes)

Key Points about Open-Ended Questions

- Open-ended questions or, as some call them, "open probes," are the most useful types in a customer environment. They tend not to mislead the customer; they enable her to speak about what is important to her.
- Open probes are a convenient way to help customers articulate their needs without too much influence on your part.
- Finally, open-ended questions are useful in the "5 Why's" environment.

Examples of Open-Ended Questions

- How's business?
- Could you give me a few examples?
- How do you use the...?
- Can you tell me all the uses of your old product?
- What problems have you experienced?
- What do you like about your current product?
- What images come to your mind when...?
- Can you tell me more?
- Can you tell me the story of when you first noticed the problem?
- Why do you think this is a problem?
- Why is that?
- What would happen if...?

Step 3: Interview Your Customers

A Selection of Various Types of Questions

Question Type	Description	Application
1. Open-Ended *(Open Probes)*	• Used to draw out a wide range of responses on a topic. • Enables your customer to get oriented and lays the foundation for a dialog. • Good to start a conversation and to explore your customer's priorities.	• How is business? • What are the most important challenges you are confronting? • What is your strategic direction for next year? • What are the problems you face when using this product? • What do you like? What don't you like?
2. Closed Probe	• Use when you require a narrow or specific answer. • Best to obtain facts and figures from your customer. • Not conducive to dialog. Can be perceived as interrogation.	• What are the dimensions of the device? • What is the temperature of the water? • Does the equipment turn on?
3. Visualizing *(Image Questions)*	• When you want to encourage your customer to speculate, without confirming your position. • Very useful to enable customers to think out of the box. • Also useful when the customer is not aware of a new technology. • Help the customer visualize a scenario	• Can you imagine if you could operate your equipment by remote control? • Suppose you did not need any documentation — would that be helpful? • Visualize a device that generates no heat. What kinds of benefits could you derive? • What if there were a product that could track the information for you? Would that be helpful? • If you had a magic wand, what would be the ideal solution to the problem?
4. Exploration	• Used to engage a customer in uncovering insights.	• Describe for me the event. • Profile the typical activity when this takes place. • Identify for me the most important activities. • When you visualize the problem, what comes to your mind? • How would you visualize the challenge?
5. Clarifying or Verifying	• Used to seek understanding or add clarity to the conversation. • This type of question should be used frequently during your dialog with the customer; it assures clarity. • Also called a paraphrasing question, as it often repeats what your customer just said.	• Am I hearing you correctly that... • From what you are telling me, I am inclined to believe that...? • I am a bit confused; can you be more specific? • Please elaborate on this subject. • Tell me more. • Run that by me once more.

Step 3: Interview Your Customers

Question Type	Description	Application
6. Fact-Finding *(similar to Closed Probe)*	• Use to discover steps in a process. • Also appropriate when helping the customer make a decision or a choice.	• What happens next? • Then what do you do? • Are you interested in heat reduction?
7. Feeling	• Used to understand the approach or how a customer feels about a subject. • Used to determine the opinion and conviction.	• Why do you feel this is the best approach? • What is your opinion on how this can best be handled?
8. Draw-in or Developmental	• Designed to draw out a broad response to a narrow topic. • Similar to an open probe or "open-ended"questions.	• Can you give me an example of what you mean by that? • Can you think of other situations similar to that? • Does this happen with all the units?
9. Third-Person	• Used to remove any guilt. • Creates a safe environment for a touchy subject.	• So, some in your industry believe this is not correct? • When I talk to others about your company, they don't seem to fully agree with you.
10. Reversing	• A powerful way to avoid giving too much information too soon. • A way to avoid leading the customer. • Also known as answering a question with a question.	• You have asked me where the menu is. Where would you expect it? • You ask, "Why do we do it this way?" How do you feel we should do it? • When you say, "We did not deliver it on time," when would you expect it?
11. End-of-Conversation	• A way to allow the customer one last opportunity to speak. • Ensuring that there is one last opportunity to capture valuable information.	• Are there any other topics we should have discussed? • Is there one message you wish to send to my company? • What one thing should I tell my team when I return?

Conclusion

The above table simply illustrates the wide choices available to interview your customers. It is not intended to be a formula, nor a cookie-cutter menu. Rather, we suggest that you become familiar with the questions and learn to use them as the situation calls for.

We all have our own styles of speaking and asking questions; we suggest that you adapt the question to suit your personality and style. Otherwise, you might come across as artificial or awkward.

Step 3: Interview Your Customers

Questions to Avoid

There are questions that are very tempting, but should be avoided because they have been proven to provide little or no valuable information.

Question Type	Description	Application
1. Leading	• Designed, purposely or not, to force a preconceived answer. • These are also known as leading the witness, from the term used in courts of law. In the colloquial, it is known as "putting words in the customer's mouth." • The answer will be biased, mostly to satisfy your own presumptions and of little value — in some cases they may be dangerous; as the information may be used to validate a preconceived assumption. • Your customer may feel insulted by biased questions, as she will assume that you already have the answer and are simply manipulating her to validate your bias.	• Don't you agree that brand X is best? • You do believe that owning this product will improve your productivity, don't you? • We are really excited about this new product. We believe it will be a real winner; what do you think?
2. Useless	• These are questions that, over the years, have been shown to provide no value.	• How much would you pay for this feature? • What features do you want in this product?
3. Questions with Jargon	• Using the terminology that lives only in your company or using TLAs (three-letter acronyms) that your customer may or may not understand. • Using difficult terms to impress the customer with your deep knowledge of the topic.	• What if the APC could connect to the F-1000 and yield an incremental cycle? • The new MABITA provides an excellent perspective on the saturation of the subtractables; what do you think?
4. Market-ese	• Using terminology for marketing and sales when none of your customers are in marketing.	• What is the addressable market share for the early adopters of the emergent technology?

Step 3: Interview Your Customers

Questions by Usage

In every interview there is a natural flow of introductions and greetings at the start of the session, followed by the explanation as to why you are there and what you intend to do the information. Once you get past the start, you need to apply a second natural flow to the topics you discuss — use this table to help you. As with all examples, this is not a complete list and is intended to help you create your own set of questions.

Usage	Description	Application
1. Starting the Interview	• Once you get past the introductions and stating the reasons for your visit, starting the interview is generally a bit awkward. You must select questions that set the stage — your customer must first get oriented. • We suggest starting all interviews with an open ended question, simply to set the direction of the conversation. • You must be prepared to quickly follow up with additional probes to get to the topics on your discussion guide.	• What keeps you awake at night? • How do you see your operation changing in the next three years? • What are the most important opportunities to improve your process? • If you could solve one problem to improve your profits, which would it be? • What is the biggest impediment to improved productivity in your operation? • Who are your most important customers? And what types of demands do they place on you?
2. Identifying the Problem	• In sales training, end users are also called problem owners because customers are always trying to solve a problem or address a challenge. • A key item in every interview is to discover the problem statement; that is, the one challenge facing your customer where you can provide value.	• Describe for me the unsolved problem? • What types of problems have you experienced? • What is the most important problem you are confronting? • Identify for me the new problem? • What disrupts your process the most?
3. Helping the Customer Visualize	• Customers, like the rest of us, are focused on their day-to-day challenges and have little time to think about their problems, let alone possible solutions. You will be faced with the need to help the customer visualize the future; i.e., how things could be better. • Visualizing is a quick way to get the customer away from the day-to-day block and into the world of possibilities. Properly used, visualizing questions will yield valuable information.	• Imagine if you could operate your equipment by remote control. • Suppose you did not need any documentation; would that be helpful? • Visualize a device that generates no heat. What benefits could you derive? • If you had a magic wand, what would be the ideal solution to the problem? • Visualize a new product in actual use. • Paint a picture of the perfect solution for me.
4. Closing the Interview	• The last few minutes of your interview are a treasure trove of information if you know how to take advantage of the time. The customer is fully engaged and clearly understands the purpose of your visit. A proper closing will enable the customer to give you valuable information.	• Anything else you wish to tell me before we bid good-bye? • What one thing do you wish for me to remember? • What is the one message from you that I must take to my team? To the president of my company? • What must we avoid doing?

Step 3: Interview Your Customers

Advanced Communication Skills

Better Interpretation, Better Understanding

Semantics, also called linguistics, is all about how people interpret the meaning of words. Said another way, there is the meaning contained in the dictionaries, and then there is the meaning interpreted by people. Semantics is about how we ascribe the meaning of words.

Bottom line, what your customer says may not be what you interpret. Worse yet, what your customer says may not be what she means. To complicate matters, consider that most of the meaning in human communication is derived from things other than the very words we use. Tone of voice, body language, context, etc. provide as much as 80 to 90 percent of the meaning. This is part of the challenge in understanding your customers.

We will overview a few critical topics in semantics and body language. We suggest that you pursue additional knowledge through books, seminars, and conferences on the subject. The better you interpret, the better you will understand your customer.

The 5 Why's

Important: The 5 Why's is the most powerful tool to uncover customer requirements

Help Customers Reflect Their True Needs

This is a simple methodology to obtain valuable information from customers. As the name implies, in the literal way, it means asking the customers "why" five times in order to get to the root message — i.e., the most fundamental message concerning the needs of the customer. This methodology originated in Japan and is typically associated with the process of identifying why failures occur. It is part of root cause analysis.

In the context of customer interviews, the 5 Why's serves the purpose of getting past the stock answer. It allows you to peel away the layers of information until you get the valuable information underlying the needs of the customer.

The 5 Why's is an excellent tool to get your customers to tell you their unarticulated or unstated needs. The process of asking why causes the customers to reflect on their actual needs.

General Rules for the 5 Why's

1. You can use as many why's as needed. While the name of the methodology is "5 Why's," it simply means that you can continue to ask why until you get useful information. In some cases, three why's will suffice, yet in other cases as many as seven will be needed. Use your judgment, and stop asking when you feel you have obtained the root message or the valuable information you are seeking.
2. Avoid the toddler's curse, when a 3-year-old child drives you crazy asking "why." Instead, let the customer talk, then ask why. You can also use other ways to ask — for example: "Tell me more" and "Why is that?"
3. Help the customer by restating her previous answers or by repeating the questions. For example: "You said that this product is difficult to use, and that this is because the menu system is not intuitive; now, can you describe why the menu system is not intuitive?
4. The intent of the "5 Why's" is to get to a level where you get "actionable input" and not generalities. Actionable input is customer information that you can actually use in developing requirements and features.
5. The "5 Why's" is not intended to be used as an interrogation tool. That is, one does not actually ask "why" five times. We encourage you to use your own style to apply the concept.
6. Your role is to enable your customer to talk and express his "root message."

Step 3: Interview Your Customers

Examples: The 5 Why's

Example 1

Q **INTERVIEWER:** How do you feel about this software product?
CUSTOMER: It is very difficult to use.

1st Why **INTERVIEWER:** Tell me more?
CUSTOMER: The menu system is not intuitive.

2nd Why **INTERVIEWER:** Tell me why it's not intuitive?
CUSTOMER: I get lost all the time.

3rd Why **INTERVIEWER:** Why do you think you get lost using the menu?
CUSTOMER: Well, there are too many layers and no way to get back.

A **ROOT MESSAGE:** The menu system has too many layers and no way to get back, which makes customers feel that the product is too difficult to use.

Note the root message — the actionable input from the customer.

Example 2

Q **INTERVIEWER:** Tell me about your printing needs.
CUSTOMER: I want a printer that prints at more than 20 pages per minute, and 30 would be better.

1st Why **INTERVIEWER:** Why is that important to you?
CUSTOMER: The one we have now does 10 pages/min and it is too slow.

2nd Why **INTERVIEWER:** In what circumstances is it too slow?
CUSTOMER: In particular when we have a queue of documents with heavy graphics.

3rd Why **INTERVIEWER:** Tell me more.
CUSTOMER: We often have to print 60-100 page reports at the last minute before a presentation, and these reports contain lots of graphics. It is not atypical to have 3-5 users printing reports at the same time.

A **INTERVIEWER:** What about documents with text only?
CUSTOMER: No problem there. It is only when we have these heavy graphics reports and when 3-5 users are queueing.

From feature to problem, don't stop probing until the customer voices the problem.

The Ladder of Inference

Description

This occurs more often than we care to admit. We witness some objective evidence, and immediately set about creating inferences, which in many cases are wrong — our minds work overtime. Worse yet, we take selected objective data and build our own story, fueled by our active imaginations. This effort generally leads to a conclusion our minds have affixed; i.e., one not based on reality. This concept was made popular by Peter Senge, but originally it is attributed to Chris Argyris.

In the example, a meeting has started. The person who called the meeting observes that one member, John, is 15 minutes late. This is the only "fact." As he moves up the ladder, the person starts to create a path toward an affixed conclusion, which most likely is not correct. For example, what if John is late because he had a flat tire? Or there was an accident on the freeway?

Applications

We must be sensitive to the difference between observable fact and fiction. It is easy for us to make inferences about customers, which may not be correct or may cause us to create the wrong product or feature.

Process

You must be aware that there is a possible trap called the Ladder of Inference. This alone will keep you from creating stories about your customer that do not exist in reality.

Use your validation and verification techniques. Ask the customer to confirm your assumptions.

Use your probing techniques, ensure that you ask enough questions to clearly understand not what the customer says, but what she actually means.

Example: The Ladder of Inference

This example shows how someone might take one observable "fact" and jump to a conclusion. Frequently we fall victim to this challenge, the tendency to "jump the gun" and reach a flawed conclusion based on an observable fact. Keep this in mind when talking to your customers: It is best to verify your conclusions to ensure clarity.

Step 3: Interview Your Customers

The Agree-Disagree Filter

Description

This is a common communication malady. It goes something like this: The customer is talking to us, and we are very busy, in our minds, agreeing and disagreeing with what the customer is saying.

Applications

We are not listening — we are judging everything that is being said and placing it into two buckets: agree and disagree. Consequently, we are wasting all the valuable information that is being given to us.

Your role in a customer interview is to internalize everything that is being said, not to judge — that comes later.

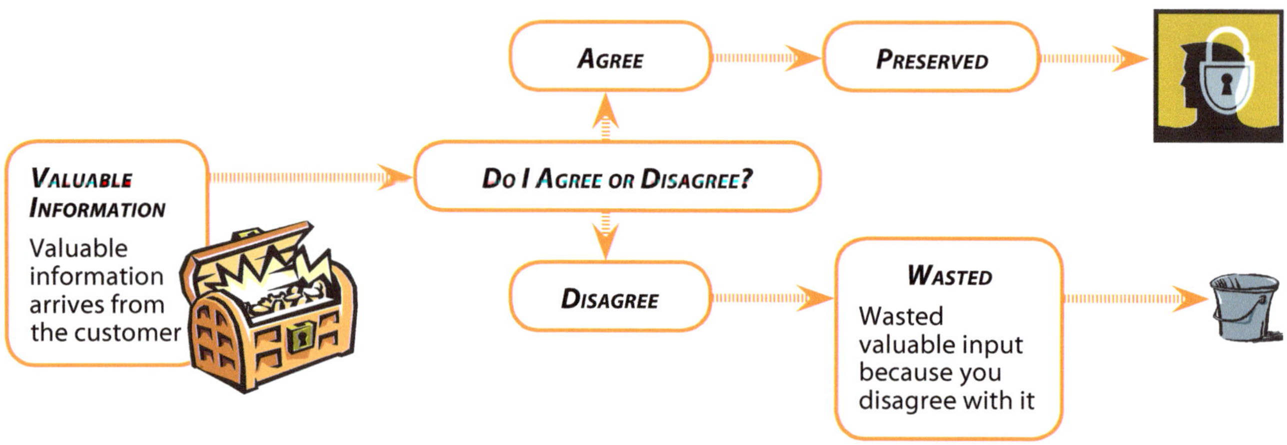

The Agree-Disagree Filter

The Agree-Disagree Filter is a bad habit; constantly evaluating the input from your customer and agreeing/ disagreeing with it. When you disagree, it is very likely that you will disregard valuable input from your customer.

You should be willing to listen to all the input; your obligation is to fully understand it. There is no room during an interview to agree or disagree; the objective is to understand the "why?"

Process

You must be aware of the filter. Knowing that is exists will help you manage it — particularly if it is a habit you have acquired over the years.

Practice active listening by immersing yourself in the conversation. Allow everything the customer is saying to enter your mind, but this is not the time to agree or disagree. Take copious notes to ensure that you are internalizing the message, and above all immerse yourself in the interview.

Use your probing techniques; ensure that you ask enough questions to clearly understand not what the customer says, but what she actually means.

The Already Thinking Challenge

Description

We all suffer from this malady. Your experiences, upbringing, morals, and even your mood impact the way you interpret messages from your customer.

Some say you color the input with your previous experiences and biases; this is, naturally, a barrier to true interaction with your customers.

For example, you go interview a customer who works for a company with which you have had a bad experience. There is the risk that the honest input from the customer will get biased by your previous experiences with that company.

Essential "There is the risk that the honest input from your customer will be biased by your Already Thinking Challenge."

Applications in Customer Interviews

You want a clean slate; that is, you want to listen to the input unfiltered by your personal bias — a tall order indeed. This is harder to do than it appears. You are biased by the culture of your company, which is natural and also by what other people in marketing say, and all the "folklore" in your company, which includes those perceptions that circulate in your organization. These biases are "already thinking," sometimes called pre-conceived notions; either way they will cloud your perception when interviewing customers, when you must have absolute objectivity.

Process

As with the other maladies, being aware that you may be prone to "coloring" input will in itself help you manage it. Make an effort not to filter input from your customer — open your ears and internalize everything that your customer says.

This problem supports the strategy of several teams going to interview customers. The other members of your organization may not have the biases you have. In addition, the layer number of interviews will tend to eliminate your biases.

The Ladder of Abstraction

The Ladder of Abstraction

The concept of the Ladder of Abstraction has its origin in the study of semantics. It was first postulated by Alfred Korzybski, but its popularity is attributed to S. I. Hayakawa in his work "Language in Thought and Action" (1990, fourth edition). It is a very powerful concept that permeates all levels of society, yet few people truly apply it. Naturally, it is critical to customer interaction.

We have all read claims in advertisements, which go something like this: "It cleans 30 percent better!" Better than what? The statement is senseless because it does not provide context or reference. Many of our communication problems stem from people using a language that contains details that are different from the level of detail of the person doing the listening.

For example, the CEO of a company speaks in terms of dollars, while a technician speaks in terms of nodes in a circuit. You can visualize that these two individuals would have difficulty communicating. The CEO is worried about the price of the stock, and the technician is worried about cross-talk in her circuit design. These two people then are said to be at two different levels of abstraction. The wider the gap, the more difficult communication becomes.

Step 3: Interview Your Customers

Example: The Ladder of Abstraction

Let's look at example that may be very relevant. Say that a customer (her name is Sue) wishes to understand the distribution of unpaid bills in her company. She obviously wants to obtain this information, in order to make an informed decision so that she can improve the profitability of her company Right away you can see the various levels, starting with wanting some information. And up to the highest level, which is improvement of profits.

Let's further say that she is talking to a computer scientist (his name is Fred). He interprets her message as wanting a table from a database listing the occurrence of unpaid bills. Further, Fred knows that in order to obtain the data, he will have to access a specific server, which is located in a different building. In this case, you can see the ladder building, but toward the tactical level (details).

The example with Sue and Fred illustrates the concept of the Ladder of Abstraction.

Notice that all the statements in the ladder are true. But at the same time, your "truth" depends on where in the ladder you are. The engineer, not surprisingly, is oriented to the tactical (detail) level. Sue, the CEO, is more toward the abstract; i.e., profitability.

You can see that Sue and Fred would have difficulty communicating, as they would ask each other different questions. They must seek common ground — which is another concept that asserts that there is a point on the ladder where the two parties can dialog productively. Another name for common ground is "the point of convergence." Obviously, it is that point where the two parties can communicate.

Finally, if the gap in the ladder is too wide the individuals will not communicate; rather, they will argue for their "truth," not knowing that they are in agreement from the start.

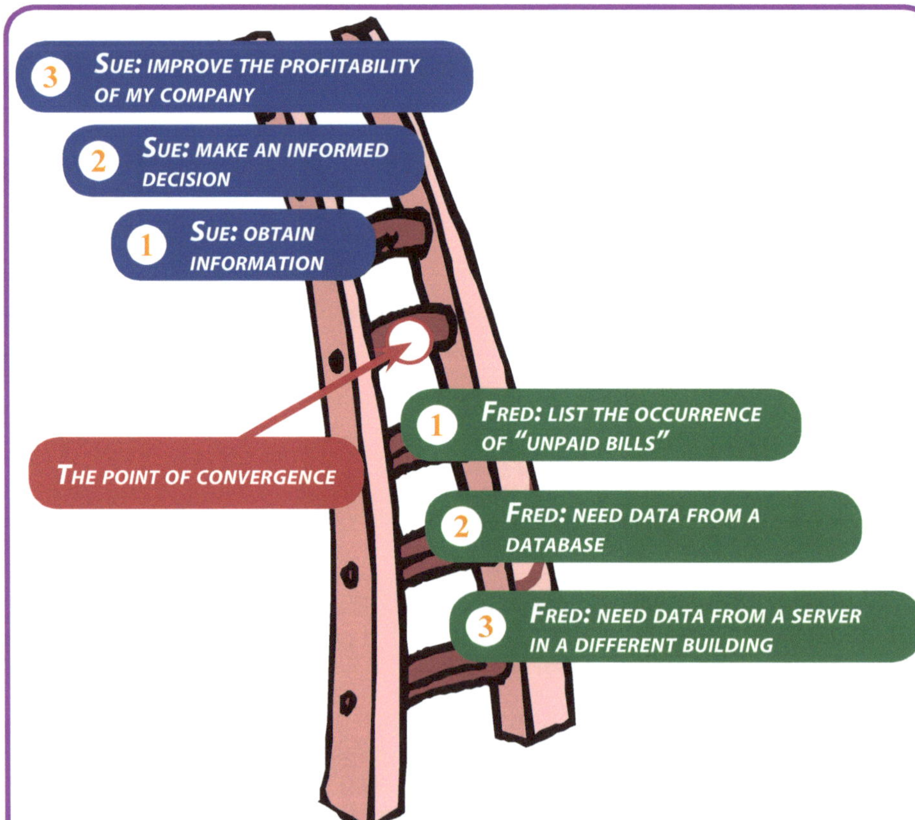

Note the "Point of Convergence," which is the point where the two parties can find common ground. Fred, the technical person, lives in the details of technology, and his personality is also driven by his passion for technology. On the other hand, Sue is a businessperson equally passionate about her field of endeavor. This gap makes it difficult for the two of them to communicate. The intent of the point of convergence is to encourage all of us to move up or down the Ladder of Abstraction in order to find common ground where we can collaborate.

Step 3: Interview Your Customers

Example 2: The Ladder of Abstraction

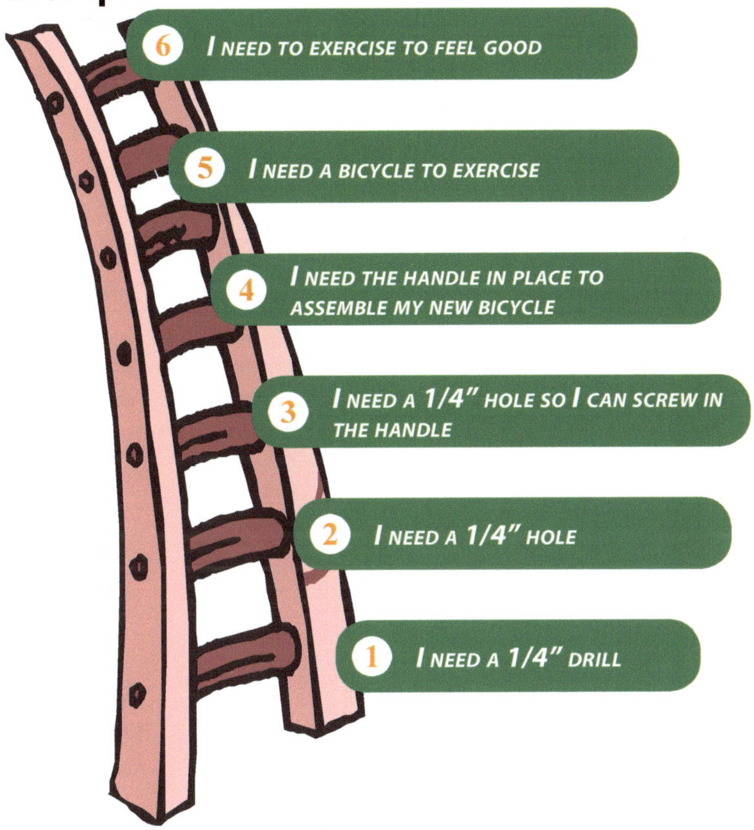

If you start reading from the bottom of the ladder toward the top, you will notice that, as you move up, the concepts become more strategic (abstract). Thus, what appears to be a mundane need for a 1/4" drill is actually driven by a very strategic need for well-being.

The rule states that all the items listed above are true at the same time; also, that things become more abstract (strategic) as you move up the ladder.

Critical Principles of the Ladder of Abstraction

- Our reality is described by levels of abstraction.
- In a Ladder of Abstraction, all levels are "true" at the same time.
- There is a point on the ladder, called "the point of convergence," where the two parties can communicate; i.e., find common ground.
- Often people argue only because they are at two different levels in a Ladder of Abstraction.

Application in Customer Interviews

Customers are looking for solutions; in reality, they don't much care about the details of how you provide the solution. This leads us to a trap in customer interviews. Many of us go in trying to find the "what"; i.e., the details. For example, "What do you measure, Mr. Customer?" We must instead focus on the "why"; i.e., "Why do you wish to measure, Mr. Customer?"

We must focus on the problem that customer is trying to solve so we can provide a solution. In other words, knowing the "why" allows us to provide elegant and valuable solutions. If come back to your office after a series of interviews and all you have is "what," you have missed the most valuable information, the "why."

Let's go back to the example of Sue and Fred. She does not want a table; she wants "information" so she can make informed decisions. Thus, if Fred knew this he could provide a graphic depiction of the unpaid bills, thus allowing Sue to be delighted as the information would be more useful than a simple table.

Step 3: Interview Your Customers

Process

Learn how to build ladders of abstraction. They are simple but provide considerable information. They also allow you to rise to whatever level your customer chooses.

When interviewing customers, the focus is on the problem she is trying to solve. The critical strategy is to focus on why your customers do what they do. Avoid arguing with your teammates when the only difference is the level of the ladder; for example, engineers and scientists will tend to be at the detail level.

The "What" and the "Why"

In the world of customer input, the "why" is always more important than the "what." Understanding the motivation (why) for the behavior of the customer is more useful than knowing "what" the customer is doing.

Thus it is always better to understand why customers do what they do. For example, you ask a customer "What are you measuring?" The answer is likely to be, "I am measuring a distance." On the other hand, if you ask, "Why are you measuring?" The answer might be, "I am measuring the distance because I am trying to save money." The motivation then, is to save money, not to make a measurement. You are now able to provide better solutions to your customer, as anything that will save her money is valuable.

The Satchmo Principle

Description

Famous musician Louis "Satchmo" Armstrong was asked how he played the trumpet so brilliantly. He responded "I don't know, I just do." This is the origin of the principle: Once you learn something, it is very difficult to feel what it was like not knowing it.

Remember when you started using the personal computer, how intimidating it was, how uncertain you were about using it. Yet, today you see the PC as a "domesticated" tool. Now, it is difficult for you to remember what it was like not knowing it. Further, you may wonder why people who are new to the PC have so much difficulty.

Application in Customer Interviews

You are very familiar with your product; to you it is very easy to use, understand, and manage. Yet your customers may find it daunting. This is because you spend many hours working, talking and thinking about your product. To your customers, it may be simply a tool they use only occasionally. Thus, it is tempting for you to assume that your customers have the same skill level with your product. In some extreme cases you might "blame" the customer for not knowing how to use your product, or for not being smart enough.

Someone said to be truly customer-driven, you must look at the world through the eyes of your customers. This is very wise, but also very difficult to do. You spend every day working on your product, but your customer does not.

Process

Be empathetic to your customers; remember that, unlike you, they do not spend the entire week thinking about your product. Be open to their suggestions, even if they seem infantile, immature, or naive. After all, your customer is always right — always. For example, if your customer says, "It is difficult to use," then it is!

The Awareness Curve

Description

There is a natural progression that all humans possess to reach commitment. It is called the Awareness Curve; it is the steps that we must go through to go from "not knowing" to "commitment."

For example, to make a commitment you must first have awareness of what you are committing to. But awareness is not enough for commitment, so you need "understanding," and so on.

As shown in figure "Awareness Curve," the full progression starts at the bottom with "no awareness," and goes all the way to "action." The world of advertising is well-informed about the Awareness Curve. They use it to cause us to purchase products. For example, you cannot buy a new type of soap if you are not aware of it. Thus, they set about creating awareness. Then, as they call it, "preference," and finally "purchase."

Application in Customer Interviews

First, you are already committed to your product. Your customer may not even be aware of it. You also understand your product very well, while your customer may not. This gap may create a barrier to communication during your interviews. Said another way, you and your customer are at two different levels of the Awareness Curve, and the larger the separation the more difficult communication may become.

The Awareness Curve also tells us about helping customers understand our questions. They (customers) need to go from "no awareness" to "action," which in this case means answering your questions. You may need to provide more context and information to enable the customer to give you an informed answer.

In short, the Awareness Curve is a wake-up call to alert us that our customers are very busy in their world. They are worried about their products and business. We, at times, assume that our customers are as committed to our products are we are, but it is not necessarily so.

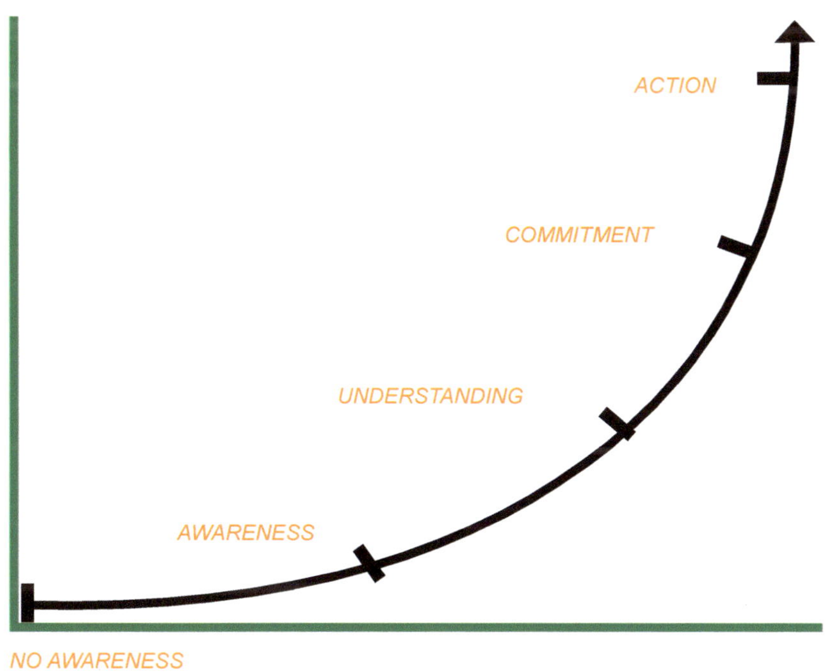

Awareness Curve

Essential

"Never accept a feature as a final answer."

PHASE II

Interpret

Step 4: Process the Information 133

Everyone who interviewed customers must now individually review his notes for relevance and then transfer the most relevant passages to yellow stickies. This process will give you invaluable perspective during the Post-Visit Meeting (PVM).

Step 5: Conduct the Post-Visit Meeting 139

The meeting (PVM) is the most important step in your entire product development cycle. This is where you determine the customer needs that you will address, which will drive the development program. Careful attention to the logistics is critical.

Step 6: Break Voices into Images and Statements 147

Use customer input to pinpoint relevant market needs. If you know what your customers need, then you will be equipped to offer products your customers truly value.

Step 7: Define and Document Requirements. 149

Define "good" requirements (and avoid "bad" requirements) and envision what features the market values.

Where You're at in the High-Level Process

— VOICE OF YOUR CUSTOMER — / — VALUE CREATION —

1 Investigate	2 Interpret	3 Innovate	4 Incorporate
STEP 1. PLAN YOUR VISIT	STEP 4. PROCESS THE INFORMATION	STEP 8. MARKETING AND ENGINEERING COLLABORATE	STEP 9. DEFINE SPECIFICATIONS
STEP 2. ORGANIZE YOUR VISITS	STEP 5. CONDUCT THE POST-VISIT MEETING		STEP 10. DEFINE ENGINEERING SPECIFICATIONS
STEP 3. INTERVIEW YOUR CUSTOMERS	STEP 6. CREATE IMAGES AND STATEMENTS		
	STEP 7. DEFINE AND DOCUMENT REQUIREMENTS		

Step 4: Process the Information

Making Sense of What Customers Said

Introduction

At this point you have completed the Investigate phase in the Master Process and are ready to initiate the Interpret phase.

Once you and your team members conclude all the interviews, you will come back from your round of customer visits with an abundance of notes (it is not uncommon to have several hundred pages of notes, video tapes, audio tapes, photos, diagrams, etc.). How do you make sense of all the things your team has heard? This is where a process is strongly needed. Without a process, the selection of these key customer needs will be open to the interpretation of each team member, and the selection of key needs will be left to the more vocal or convincing. You have invested time and energy to get all this customer input. You need to put a strong emphasis on making sure that all relevant information is taken into consideration.

This section presents a process to organize the information you have heard and documented into a coherent representation of the customer problems, taking all inputs into consideration. It is a team process, and it takes time. Your team should plan extended meetings to fully analyze and understand your customers' messages. But the results are worth it, by the quality of the resulting understanding and the building of a strong consensus among the product team on what the customers said.

Important: This detailed process for the "Post-Visit Roadmap" is made up of 18 major steps. For your convenience, we have divided it into three sections.

The Post-Visit Roadmap

Once all the interviews have been done and the team is back in the office, it's time to start making sense of what the customers said. The first step on the Post-Visit Roadmap is for everyone to consolidate all their information. After the consolidation, the process moves to the Post-Visit Meeting (PVM).

Essential: "It is important that each person captures the message from the customers from his or her own perspective."

The PVM's role is to take all that information and distill it into a shorter set of customer voices and images that represent the key topics raised by the customers. The PVM is an extended meeting, sometimes several days long, but seldom less than one day for medium-size projects. Do not short-change the PVM; allow for ample time, as it will lead you to defining the requirements of your new product. And bad requirements are a key reason for new-product failures, so why take the risk? You will face some resistance to this meeting; people are anxious to start developing the new product, and they believe they have all the information they need. As a product leader, your tenacity and leadership will win the day.

Important: This process begins back at the office, when all the visits have been completed.

These steps must be done by all those who made customer visits; make certain that everyone understands his or her responsibility.

Consolidating Your Notes

Essential Ingredients for Consolidating Your Notes

To a large extent, this process starts with taking good notes during each interview so you have a solid foundation for consolidation. As you recall, this entire methodology rests on the assumption that everyone who participated in the interviews must take careful notes.

The consolidation must be done by each individual, prior to the Post-Visit Meeting. Thus, if you had three teams made up of two people each, the consolidation of the notes must be done by the six individuals, separate and independent of each other. This is one place where collaboration is not desirable. It is important that each

Step 4: Process the Information

person captures the message from the customers from his or her own perspective, without influences such as other team members.

Consolidation is made up of two distinct steps. First you must review your notes and as you proceed, start highlighting passages, phrases and other information of relevance (use a standard yellow highlighter for this step). What is relevance? Use your judgment; you know the objectives of your set of visits and you understand the product you are developing, so therefore your judgment is valid. At times, you may need to review your notes more than once, and as you review, more information will become relevant. At the end of this step, you will have highlighted a considerable number of passages.

Step number two of consolidation is transferring most of your highlighted passages to yellow stickies. Here again, your judgment is important. You start by reviewing all the highlighted passages from the previous step, and start writing the more important ones on yellow stickies. What is more important? Again, use your judgment. Much like the highlighting step, you have an understanding of the objectives of your customer visits and the product you are trying to develop, which gives you the needed perspective to make the selection.

At the end of this step, you will have a number of yellow stickies, perhaps as many as 50 or even 100. There is no rule of thumb as to the right number. You are now ready to go into the Post-Visit Meeting (PVM).

The overall strategy to process the information from your interviews can be seen as a funnel with filters. When you conclude all your interviews, you will have a considerable amount of information. Consider all the notes, photos and audio recordings that you and your team obtained. The approach is to take all this information and gradually reduce it to a short set of customer requirements.

Step 4: Process the Information

The Strategy for Consolidation

At the conclusion of a set of customer interviews, you and your team will have a considerable amount of information. The challenge is to separate the valuable information from the trivial. Consider that you are looking for a minuscule piece of insight that will enable you to create a winning product; that is, the information from your customer that will add special value to your product.

For example, say that you defined a set of customer visits made up of 17 customers. Also, say that you sent three teams of two people each to perform the interviews. Also assume that each person captured five pages of notes. As you can see, you end up with a considerable amount of information, which you must now prioritize.

The strategy is to adhere to a process where you begin to discard information that is not important, but at the same time avoid throwing away information that is. This is the strategy behind the process of Interpreting and Innovating.

Additional Tools

Summary

The Critical Tools section of this guidebook, details several tools that will help you during the Interpreting Phase, especially when processing the information and making sound decisions.

See Also Critical Tools

Yellow Sticky Protocol

The Yellow Sticky method allows you to take massive amounts of customer data and break it down into bite-sized pieces. It is an efficient and effective tool that ensures group participation and accelerated documentation.

See Also Yellow Sticky Protocol

Affinity Diagrams

We are not linear thinkers. We tend to jump around when brainstorming. Using Affinity Diagrams is a simple method for organizing the brainstorming session — the non-linear flow of ideas — into useful, documented facts. And it is a fine way to ensure consensus.

See Also Affinity Diagrams

Prioritizing Affinity Diagrams

Prioritizing is all about making decisions and reaching consensus. The secrets are selecting the right criteria to use in making those decisions and clear communication. Use the same objective set for the Affinity Diagram to set the method for prioritizing.

See Also Prioritizing Affinity Diagrams

Karnaugh Maps

A 2x2 matrix, or Karnaugh Map, is a participative tool that enables a team to prioritize or select items from a group of many, using a set of criteria. This simple mathematical tool enables you to make better and faster decisions.

See Also Karnaugh Maps

Step 4: Process the Information

Level 2 Main Process: Consolidating Your Notes

Important — Prepare Ahead

Start thinking about the extended meeting in this process. For example, schedule the date, venue and length now. Make this process an assignment for each person who had contact with the customers. Provide a set of guidelines, such as the Yellow Sticky Protocol. Ensure clarity and a firm due date.

9

Agree on the Yellow Sticky Protocol

You and your team must be in consensus about the yellow sticky protocol you will use to help you track the information.

See Also — Yellow Sticky Protocol

Important — No Collaboration for the Next Three Steps.

These next three step should be done individually, even if two or more people visited the same customer — no collaboration, no team.

Time Frame — This review may take several hours for you and your team members.

10

Consolidate Your Notes

Organize notes from all the visits. Carefully review all notes in preparation for the next step.

Review the transcripts and audio/video recordings from your interviews.

Important — This is where audio-taping — or better yet, video-taping — pays off. This way, you have a complete record of your interview, including the tone used by your customer.

11

Highlight Your Notes

After you have organized your notes, highlight important passages:

- Read through the verbatim interview notes. Use a yellow highlighter and highlight any statement, passage, finding, statement or other information of relevance. Use your own criteria to define "relevance."
- Review your highlighted notes, and eliminate those that are clearly not relevant.
- Review your highlighted notes and eliminate the obvious duplicates.
- Do this for every interview; for example, if you do six interviews, you will do this step for all six.

See Also — Critical Tools

Important — See the Critical Tools section of this guidebook for instructions on the protocol for your yellow stickies.

You and your team will generate as many as 300 yellow stickies — you need a protocol to keep track of the information and to organize it. Adapt the protocol to suit the culture of your team.

Continued

Level 2 Main Process Cont'd.

Cont'd

12

Create the Yellow Stickies

After you have organized your notes and have highlighted the important passages:

- Read all the passages (sections) which you highlighted.
- Transfer as many of the highlighted passages to yellow stickies as necessary, you make this decision. You know the objectives of your visit program.
- **NOT ALL THE HIGHLIGHTED PASSAGES NEED TO BE TRANSFERRED TO YELLOW STICKIES.**
- Use the protocol that you and your team have agreed for documenting the yellow stickies.
- At the end of this step, you will have a stack of yellow stickies, and are ready to attend the meeting.

Important Notes:

- Use a Sharpie marker or similar marker to write on the yellow stickies.
- You must use the protocol agreed for the yellow stickies (please see next page).
- Quotes are quotes, which means they must be transcribed verbatim from your customers. You cannot make a quote from your notes.

Important

You should now have several — perhaps 100 or more — of yellow stickies similar to the example. Your team members should also have a similar quantity.

Note that each yellow sticky has a set of codes that you and your team have agreed are important.

Example

Ease of use H Design engineer

"More than three layers of menus make products difficult to use."

Fred T Page 1, 2nd interview

Critical Information

Important

The process to obtain customer requirements will necessitate an Affinity Diagram. You or someone else must be able to facilitate the affinity process for the Post Visit Meeting. This is one step that **CANNOT BE AVOIDED**.

See Also Affinity Diagrams

Continued

Step 4: Process the Information

Level 2 Main Process Cont'd.

WHERE YOU'RE AT

At this point you and your team have:

- Organized your notes, audio, video, transcripts and other information from all the interviews.
- Highlighted and annotated these notes and also reduce the amount of data, starting the process of refining the information.
- Created a set of yellow stickies using an agreed protocol.
- Become ready to attend the Post Visit Meeting. Each person will have about 100 yellow stickies ready. Some may have many more.

CONT'D

13

SCHEDULE THE POST VISIT MEETING (PVM)

This is the most important step in the entire process, so give it all the attention and commitment it deserves!

This meeting may take a full day or even longer. Do not short-change the time.

Ensure that all the people who went on the customer visits attend this meeting.

Ensure a venue that is large enough to accommodate your team, and also to allow for considerable work on the walls — i.e., flip charts and yellow stickies.

Ensure that everyone brings their interview notes and their yellow stickies. This is critical — no exceptions as you cannot proceed without the notes.

Some teams choose to break this meeting into several shorter meetings. This is acceptable, but we suggest that the shortest meeting be at least three hours long to allow for proper interaction and concentration.

Important

FAQ

Q. Can I invite people who did not interview customers to the meeting?

A. No, we discourage it. But we also recognize that there may be occasions where it is important. In such cases, there is a protocol. Those who are guests must be observers. They cannot be active participants, as they do not have the full context of having done the interviews.

CONT'D IN STEP 5: CONDUCT THE POST-VISIT MEETING

Step 5: Conduct the Post-Visit Meeting

Introduction

Once everyone has consolidated their notes and prepared their yellow stickies, you are ready for the Post-Visit Meeting (PVM).

The PVM is an extended meeting that must be attended by all those who participated in customer interviews. We recommend that you plan for a full day, perhaps even longer. Time is needed to organize the information, but most importantly to create the right dialog in order to define the message from your customer.

It is also a highly collaborative meeting, where everyone is focused on understanding the message of the customer

Essentials for a Successful Post-Visit Meeting (PVM)

- Everyone on the team must consolidate his or her notes prior to the meeting. This is a critical step, and you must go out of your way to ensure compliance. This encompasses steps 1, 2 and 3 in the process map.
- Step 2 in the process map is equally important (highlight your notes). Everyone must create a set of yellow stickies using a pre-defined protocol. See *Yellow Sticky Protocol*.
- **THE MEETING (PVM) IS THE MOST IMPORTANT STEP IN YOUR ENTIRE PRODUCT DEVELOPMENT CYCLE.** This is where you determine the customer needs you will address, which will drive the development program. Consequently, careful attention to the logistics is critical — no loose ends!
- Their meetings will be extensive, from one to three days. Do not underestimate the duration; it's a small price to pay for such valuable information.
- We highly recommend retaining the services of a facilitator, someone with skill and experience working with teams. Some companies have internal facilitators who might be able to help.
- Obviously, you need to enforce a meeting protocol. For example, turn off all cell phones and pagers. Also, "one-break all-break." Use the silence rule when appropriate — for example, when doing the Affinity Diagram.

Essential

"The Post-Visit Meeting (PVM) is the most important step in your entire product development cycle."

Step 5: Conduct the Post-Visit Meeting

The Journey from Voices to Customer Requirements to Engineering Specifications

Everything starts with the voices you captured during your interviews (Investigate phase). Your challenge then becomes to translate the voices of the customer into customer requirements (see MRD for more details on customer requirements). You must have a set of prioritized requirements in order to innovate! From customer requirements, you develop features and attributes, which ultimately lead to the engineering specifications for your product. The final step in the journey is incorporating the innovative solutions into your new product. This is the sequence!

The Sequence of Steps

1. CUSTOMER VOICES: IMAGES AND STATEMENTS

The intended sequence that will take you from the voice of your customer to engineering specifications starts with images and statements, which are the result of your Affinity Diagram. Expressions that customers use to articulate their needs are called voices, which can be broken down into images and statements. An image is a sentence that evokes a vivid mental picture and describes the problem, challenge or issue confronting your customers. Statements are more factual and usually relate to the product rather than the customer. The best images and statements are actual quotes — a verbatim phrase, sentence or narrative — from the customers that express a clear problem or need.

2. CRAFT REQUIREMENTS

Once you have affinity, you and your team will craft the requirements — which are not features and attributes. A requirement is sentence that describes the quality (what the product shall have) and functionality (what the user or the product shall be able to do) that must be addressed in order to meet the need expressed in the image and associated statements. A requirement expresses what the offering should do — but not how it will be done.

The role of marketing is to deliver a clear set of requirements that truly represent the voice of the customer. Rigorous dialog and a robust process are needed to understand, phrase and prioritize customer requirements. Given such a challenge, it becomes clear that the Post Visit Meeting (PVM) is a critical step. At the end of the PVM you have a set of prioritized customer requirements, which is one step away from the voices.

Essential

"The role of marketing is to deliver clear requirements that truly represent the voice of the customer."

3. FEATURES AND ATTRIBUTES:

Only after you have crafted the requirements, can you and your team proceed to develop the features and attributes that will address how to solve them. Features refer to specific functions and qualities. Attributes are the actual traits of the product, what a feature is or how the feature is implemented.

4. CONCEPTS

The sum of the features and attributes that will make a complete product. There might be several concepts for the same sets of requirement as there are several ways to address customer needs. The customer will be the judge of which one will win.

5. ENGINEERING SPECIFICATION

Describe the features and attribute in terms that an engineering organization can use to design them. Use unambiguous metrics. For example: Weight, 5 pounds; size, 8x11 inches. Engineering then continues the process with technical documentation.

Step 5: Conduct the Post-Visit Meeting

Example: The Ladder

1 IMAGES AND STATEMENTS:

"When I travel by plane for a one- or two-day business trip, I carry my business documents and personal belongings, plus my laptop, which means that I have at least two bags, so I have to check one as luggage. I wish I had a single bag that I could carry on board. That would save me time and aggravation."

2 REQUIREMENTS:

The user shall be able to carry her personal belongings, laptop and business documents needed for a two-day business trip in a single carry-on bag.

The bag shall allow the owner to quickly get the laptop out without disturbing the other items for easy check through airport security.

Find more...

3 FEATURES AND ATTRIBUTES:

Bag has separate compartments for business documents, laptop and personal belongings

Bag meets airline regulations

Rolling wheels

Padded computer compartment

Pockets for pens and other common business tools in business compartment

Removable pocket for common toiletry

Find more...

4 CONCEPTS:

Carry-on bag — It is the name that will designate the product.

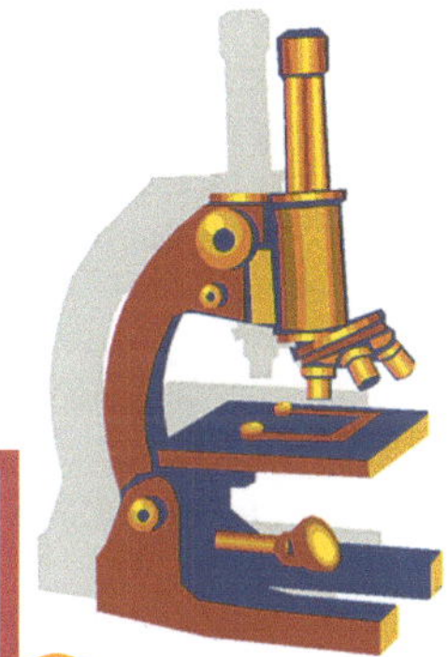

5 ENGINEERING SPECIFICATION:

Three-zippered compartments made of Kevlar material.

Find more....

This example illustrates the progression from the Voice of the Customer all the way to Engineering Specifications. Engineering will continue beyond #5 with appropriate technical documentation. The role of marketing is deliver a clear set of requirements that truly represent the Voice of the Customer.

Step 5: Conduct the Post-Visit Meeting

Level 2 Main Process: Begin the Post-Visit Meeting

Continued from Step 4: Process the Information

CONT'D

14

HOLD THE MEETING

- Start the meeting by reviewing the business objectives for your new product.
- Ask each person to read and post all the yellow stickies on a large surface, regardless of their content, validity or relevance. (This may take two or more hours.) It is essential to creating understanding and depth!
- Place the yellow stickies at random. It is not rare to end up with as many as 300 yellow stickies — fear not!
- Participants can ask questions for clarification. This is not the time to argue or debate.

Important

Allow Enough Time

These meetings are very important, and absolutely essential to the process. They generally last from four hours all the way to several days. It is not possible to compress the amount of dialog, introspection and processing of the information. Please, do not shortchange this important dialog; it holds the key to innovation and competitive advantages, but it does take time; in some cases days!

Important

Earlier in this chapter, we described the strategy to process all the yellow stickies.

The intent is to narrow the number of stickies to a manageable quantity.

Each step in the process is intended to reduce the number, but not the quality of the information.

Steps 14 and 15 are part of the strategy of consolidation.

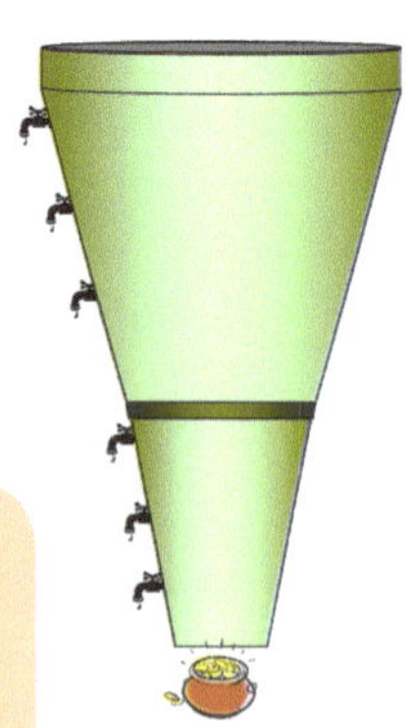

See Also
Strategy for Consolidation

CONTINUED

Level 2 Main Process: Begin Processing Customer Input

CONT'D

15

SCRUB 'N SCOUR

Scrubbing is a first pass at consolidating your yellow stickies. **IT IS NOT AN AFFINITY PROCESS YET.**

- Identify clear duplicates, but ensure consensus from the group that duplicates are indeed duplicates.
- Identify items that seem irrelevant or not important.
- Remove the duplicate and irrelevant yellow stickies.

Flexibility

Optional Step

This next step is optional. If you did not achieve a significant reduction in the number of yellow stickies in the previous steps, then you should perform a "Pick'n Choose"

You should strive to have fewer than 200 yellow stickies when you are done with this phase.

16

PICK 'N CHOOSE

Move all the yellow stickies left after the scrubbing to a different surface, such as the opposite wall. This step is important — during this process, the group will notice yellow stickies that are clearly trivial or irrelevant. Leave these items (do not move them to the new surface).

The move is done in silence. Everyone can choose to move or not to move any of the yellow stickies.

When finished, ensure consensus that the yellow stickies that were moved needed to be moved, and those left behind should indeed have been left.

17

SEPARATE VOICES FROM SPECS

After you have concluded moving the yellow stickies from one wall to another, you will still have a significant number, perhaps several hundred.

In this step, you and your team should examine and remove all yellow stickies that only have technical information; for example, specifications. Also remove all yellow stickies that have only a description of a new feature.

The yellow stickies remaining should only have voices — anything that expresses a customer need or desired outcome. In short, keep the yellow stickies that express the "why", and remove — but do not discard — the yellow stickies that only express the "what."

Do not discard the yellow stickies you remove because they will be used later in the process, after the Affinity Diagram.

WHERE YOU'RE AT

At this point you and your team have:

- Read all their yellow stickies aloud.
- Placed all the yellow stickies on a wall, perhaps several hundred.
- Started the process to reduce the number of yellow stickies by doing the "Scrub'n Scour", "Pick'n Choose" and also separated "voices from specs."
- Established a manageable number of yellow stickies on the wall — strive to have around 100 to 200 in total. If not, continue to reduce the number by repeating some of the previous steps.

CONT'D: STEP 6: BREAK VOICES INTO IMAGES AND STATEMENTS

Step 5: Conduct the Post-Visit Meeting

Examples: From Requirements to Features and Attributes

Example 1: The Process of Imaging

See Also Yellow Sticky Protocol

Easy to carry around Businesswoman

The laptop I have is not convenient. I carry it all day from customer to customer in its carry-on bag. But I also have to carry the electronic projector and my business files. When the day is over, my shoulder hurts.

The three headers (categories) are combined into a single image (message). The image is a customer quote that captures the key message from the customer.

Headers from the Affinity Diagram capture the key message of the clustered yellow stickies.

Easy on the shoulders	The laptop shall be easy to carry by an average businesswoman	The user shall be able to go places carrying all the required equipment using only one hand
Easy on shoulders Busnwmn. 42y. Once you add the accessories, the additional battery and other stuffs you get easily to 7 lbs. That's too much to carry all day.	Easy to carry Insurance agent 52y. The bag I have is bulky. Why don't they make slim ones? That's easier to carry.	Convenience Saleswoman 35y. I do not need one bag for each thing I need to carry. Carrying more than one bag is a pain. Although sometimes I only need the laptop.
Easy on shoulders Saleswoman 35y. I do not know how much it should weigh, but I know that I can carry a few books without too much pain.	Easy to carry Businesswoman 42y. When I travel by air, I have one of these suitcases with wheels. That's the most important woman-friendly innovation of the 20th century. (Laughter)	Convenience Insurance agent 52y. I'm always worried if I have all the needed accessories. I stop in so many places that I always run the risk to forget something.
Easy on shoulders Saleswoman 35y. These shoulder straps are hurting. You don't see kids carrying their bag this way anymore. If it's not good for kids, why is it for us?	Easy to carry Saleswoman 35y. I am not tall (5'). These laptops are made for tall men.	Convenience Saleswoman 35y. I do not like arriving and being unable to greet a customer because I have to carry three bags.

Adhere to your Yellow Sticky Protocol. In this case, the upper right corner captures the source of the quote or information, and everyone in the team must use the same corner for the source.

Step 5: Conduct the Post-Visit Meeting

Example 2: The Process of Imaging

In this example, the upper left corner captures the topic from the discussion that gave rise to the quote or information.

The upper right corner captures the title of the person who was interviewed.

The lower left corner captures the name of the team member who wrote the yellow sticky.

Every team member must use the same protocol.

Ease of Use — Design Engineer
"When I travel for a day or two, I carry my business documents and personal belongings plus my laptop. I wish I had one item that I could carry on board that would save me time and aggravation."
Fred T.

The user shall be able to carry business files, laptop and personal belongings in a single carry-on bag.

All in one — Design Engineer
I can only take one carry-on piece in the cabin on many flights.
Fred T

All in one — Eng. Manager
I hate to wait for my bags when I arrive at my destination. I always wonder if they are going to arrive with me, and sometimes they don't.
Roger F.

All in one — Technician
I attached the laptop bag to my carry-on case to get past the limit. Sometimes they let me get away with it, but if not, I have to check it in.
Fred T

All in one — Technician
Registering luggage and waiting for it adds precious time to my trip. I could do without this burden
Fred T

The bag shall withstand regular airline transportation "mistreatments"

Sturdy — Technician
I'm always anxious to see if my bags were damaged during travel — that's why I don't like to check them in.
MC

Sturdy — Technician
My laptop needs to be protected inside the bag. The last thing I want to see is a damaged laptop when I get to my destination.
MC

Sturdy — Mechanical Eng.
The last piece of luggage I had did not even last two years. The fabric was torn in several spots.
Roger F.

The bag shall allow easy access to the three types of goods that it contains. In particular, the laptop shall be easily pulled out from the luggage compartment.

Convenience — Design Eng
I have to take the bag out of the luggage compartment to take out my laptop.
Fred T

Convenience — Technician
I like the wheels and the handle that my carry-on luggage has.
MC

Convenience — Manager
I don't want to put everything together in a single bag. I want to have separate compartments to pack various items.
Fred T

Convenience — Manager
When I go on a business trip, I take my laptop and all my business files with me.
Fred T

Step 5: Conduct the Post-Visit Meeting

Example 3: The Process of Imaging

Easy to carry aroundBusinesswoman

The laptop I have is not convenient. I carry it all day from customer to customer in its carry-on bag. But I also have to carry the electronic projector and my business files. When the day is over, my shoulder hurts.

Easy to Learn	Easy to Use	Security
Intuitive (easy to learn) It does not follow the generally accepted practices for designating menus such as Microsoft. I had to relearn a new GUI.	Easy to use It does things on its own that I don't understand, and don't know how to avoid. For example...	Security It does not seem to be compatible with my firewall.
Intuitive (easy to learn) The Help menu is of little help. It does not allow me to solve a problem — just shows how to use each function.	Easy to use Once I've learned a new product, I do expect to use it without surprise. Your is full of surprises.	Security Setting the four Privacy parameter is a chore. You should have a Wizard.
Intuitive (easy to learn) I do not want to have to read a manual to start operating new software. I want it up and running in a few minutes. Later, when I get a bit more experience, I might want to look in the manual.	Easy to use Exporting a file to MS Word takes four steps. It is a task I (and I suppose others) do frequently.	Security It does not seem to be compatible with my Norton firewall.
Intuitive (easy to learn) When I received the package, I looked for a quick-step guide. I could not find one. I had to go through the lengthy explanations in the manual.	Easy to use I get at least two error messages per hour of operation. And usually, I have no clue what I did wrong.	Security It slows down significantly when set at maximum protection. That is just not acceptable, and in addition it is not even mentioned anywhere.

In this example, two similar yellow stickies appear in one cluster, which is acceptable as long as it makes sense to your team.

For example, use duplicate yellow stickies for emphasis or because, although similar, they convey different messages.

Step 6: Break Voices into Images and Statements

Introduction

The Affinity Diagram process is a way to cluster or arrange seemingly disparate ideas into a logical pattern. As described in the Affinity Diagram process, the yellow stickies must be arranged by the team members into logical categories; it is a form of consensus decision-making. In the case of customer visits, there is a special case of Affinity, which deserves much attention.

Important Information about Affinity Diagrams

If you have not lead an affinity diagram before or if you are not familiar with the methodology, then read the section on Affinity Diagram in the Critical Tools section of this guide.

You must become familiar with the Affinity process prior to the start of the Post Visit Meeting; otherwise, you risk a severe slowdown of the process or, worse yet, total failure.

Perhaps, there is someone in your company who can act as a facilitator for this part of the process. You can also hire a facilitator who has proven experience with this methodology.

The Purpose of an Affinity Diagram

The affinity process will help you organize the information so that you can extract the customer requirements. Once you have filtered all the yellow stickies, you will still have a large number left on the wall. At the end of the process you will have clustered all the yellow stickies into a small number of categories (headers); we suggest that you strive to end with five to seven categories. From each of the clusters you will extract a customer requirement.

Definition — Image

The customer's message as a vivid mental picture that describes the problem, challenge or issue confronting your customers.

Images and Statements

In the case of customer input, the headers will most likely be customer images because they convey a more complete level of needs. It is the representation of how it feels to be a customer — "feels" being the operative word. These images must be actual quotes from the customers, extracted from your notes and those of the other team members.

See Also — Affinity Diagrams

Images are qualitative statements that create a vision of the world, a description of need or a description of a problem — but as the customer sees it. For example:

- "I feel overwhelmed by the menus; it is like a thousand choices all at once."
  In this example, it is not difficult to visualize the frustration of the customer when dealing with the menu.
- "When I travel by air, I am always anxious to know if my checked-luggage arrived, and that it arrived undamaged."

On the other hand, voices like "I would like to use my PDA in the dark," or "I want a lightweight laptop" are not images; we call them "statements." In such statements the customers provide a solution to a problem that has not been stated. Good interviewers have learned to not accept these kinds of statements as final answers. You will see later how when we examine follow-up questions.

Images, therefore, by being visualizations of need as expressed by the customer, either positive or negative, and in the language of the customer, carry more weight than simple statements. In addition, they are usually not subject to argument as it express a customer "fact." Statements on the other hands are usually open to argument because they are not supported by any reason, and may or may not be the best answer to the customer's unstated problem.

Q: What if we do not have a quote that captures the cluster?

A: It is an opportunity for improvement, as quotes from the customers should abound. But, in a rare case, you may have to create an image. The team can visualize what it feels like to be a customer, and state the feeling in a way that captures the message.

Step 6: Break Voices into Images and Statements

From Images and Statements to Requirements to Features

Step 1: Break Voices into Statements and Images

The goal of customer interviews is to generate the maximum number of high-value inputs from the customer. These inputs will shed a light on his conscious and unconscious needs. A verbatim statement of the customer is what is called a "voice." But not all voices carry the same power and use.

For example, a customer might say: "The laptop I have is too heavy," or she might say, "I carry this laptop all day from customer to customer. When the day is over, my shoulder hurts. I wish I had a laptop I could carry as conveniently as a book." The first voice is interesting but low-level, because you do not know "why" it is bothering the customer. "Why" it is bothering the customer might lead you to innovative solutions. In the example above, the weight may not be the only reason the customer complains; the shoulder strap or the way she carries it might be cause. So the "why" is what you typically want to understand. But we get a lot of low-level voice in interviews; they are useful and they should not be neglected. They should not be taken for granted, and should be supported by an image to make sure that you get the complete picture. It is a good idea for product teams to evaluate what percentage of images they get from a round of interviews and to track their progress over time. Inexperienced product teams may not get more than five percent of images. So, out of 200 voices, they will get only 10 images. This is insufficient to really grasp the richness of the customer problems. It is difficult to say what a good percentage is, as the quality of images is also to be considered, but if you get 20 percent out of a few hundred voices, you will get 40 image — and you should be able to get real gems to help you define requirements for defining a killer product. We suggest therefore that you keep track of the number of images and that you determine ahead of time the number of interviews you will conduct (junior teams should conduct more interviews than experienced teams).

An image is what you could have gathered if the interviewer had simply asked the customer who was quoted of the first voice, "Tell me more," or "Why is it a problem?" This is also an argument for having at least two people carrying the interview as the interviewee might miss the opportunity to ask follow-up questions. We cannot insist enough on the importance of this skills.

So, when processing interview voices, break them in two categories: the statement voices (S voices) and the image voices (I voices). It helps to identify each of them with different colors (a green dot for images, for example) and to group them together to give a full picture of the problems expressed by the customers. "I voices" typically end up at the top in the affinity diagram.

Step 2: Transform Voices into Requirements

The next step is to transform these voices into requirements. Customer requirements express what the **CUSTOMER/USER** shall be able to obtain/do/have, or what the **PRODUCT** shall be able to do/have in order to satisfy the customer needs. Typically, an image and associated statements generate several requirements, while a single statement voice generates only one or two.

Let's take the following customer image: "I am a professional account manager. I carry this laptop all day, from customer to customer, doing presentations around the country. When the day is over, my shoulder hurts. I wish I had a laptop I could carry as conveniently as a book." Here is a sampling of requirements into which this customer image could be transformed:

- A typical business-woman shall be able to carry the laptop all day without fatigue.
- The laptop shall be as easy to carry as a book.
- The laptop generates a feeling of competence to its user.
- The laptop shall handle shocks and hazards typically encountered in air and ground transportation.

Important: If you do not have images to fit a particular need or problem area, create one that the product team can agree on by visualizing yourself in the customer's world and doing the activities she does.

Step 7: Define and Document Requirements

Good Requirements

Introduction

Expressing good requirements is one of the most important steps in new product development. Actually, it has been shown that "bad" requirements are the key reason for product failures. But to get started, what is a requirement? And what is a good requirement? And how do you know if you have a good set of requirements? It is important that the product team has a common answer to these questions

Customer Requirements

Customer requirements express what the customer/user shall be able to obtain/do/have, or what the product shall be able to do/have in order to satisfy customer needs. Requirements ARE NOT descriptions of HOW a product will satisfy customer needs. The HOWs are solutions. Expressing customer requirements is the responsibility of marketers and is generally documented in a Market Requirement Document (MRD). From customer requirements the team then looks for possible solutions that will satisfy these requirements (this is the Innovate phase). Too often product teams jump directly from customer needs to product features and even specifications without expressing clearly the customer requirements — and that is an error. Bypassing this step hinders the team from finding innovative solutions to market problems.

Examples of a Customer Requirement

CUSTOMER REQUIREMENT: "The user shall be able to comfortably work on her laptop in an airplane coach seat for five hours."

This requirement can also be expressed as a product requirement, which is the passive-tense version.

CUSTOMER REQUIREMENT: "The laptop shall be able to be operated comfortably for up to five hours in an airplane by a user in a coach seat."

Notice that these requirements do not mention how the product shall address the customer needs. This is the role of solutions.

Examples of Product Requirements and Solutions

Below are a mix of requirements and solutions. Try to determine which is which.

- The battery of the laptop shall have a minimum life of five hours.
- The laptop shall contain a spare battery so that it can be operated for five hours with battery power.
- The user shall be able to plug the laptop into a reliable source of power close to his seat.
- The laptop shall be no larger than 11.5" x 17" (to fit on a coach tray table).
- The laptop shall pass the security check without physical damage.
- The screen shall have side blinders (to not be viewed by neighbors).
- The laptop keyboard shall incline so that it is comfortable when used on the tray table or lap.
- The laptop base shall not get hotter than 55°F (to be operated comfortably on the user's lap).

Please note the way requirements are expressed with a short and clear sentence containing the verb "shall." Adopting a structured way to express requirements is a discipline that will go a long way in facilitating understanding between engineering and marketing.

Types of Requirements

There are several types of requirements to consider. First, a particular product will generally have several stakeholders that will be involved at the various stages of the product life-cycle: buying, installation, use, maintenance, disposal and overall results. Each of these stakeholders has requirements that need to be elicited and documented. When expressed as a product requirement, there are at a minimum four types of requirements to consider.

Business Requirements

Business requirements relate to the business results the customers want to obtain. Going back to the five categories of benefits and costs, these business requirements deal mainly with the operational and economic benefits, and most of the costs categories. Examples:

- "The customer shall be able to increase yield by 2 percent minimum."
- "The product shall allow the customer to increase yield by 2 percent."

Functional Requirements

Functional requirements relate to what the product shall **DO**. Examples:

- "The product shall warn the user when the battery reaches a low level."
- "The product shall play MP3 songs and DVDs without booting up."

Non-Functional Requirement

Non-functional requirements relate to a quality the product shall **HAVE**. Example:

- "The product shall be ready to use by people of age 65-70 within 30 minutes of unpacking."

The most-frequently required non-functional requirements include:

- Look and feel
- Usability
- Performance
- Operation
- Maintainability
- Portability
- Security
- Cultural and political
- Legal

Constraints

Constraints restrict the options available to the development team. Examples:

- "The product shall have a manufacturing cost of $1,200."
- "The product shall interface with an IBM 567 server."
- "The product shall be used by all teenagers, male or female."

Step 7: Define and Document Requirements

Characteristics of Good Requirements

COMPLETE: A requirement must fully describe what the user shall be able to do or experience.

TRACEABLE: Each requirement should be traceable to a real customer need, and that need should be documented somewhere.

FOCUSSES: Each requirement should express a single need

VERIFIABLE: The product team should be able to verify if the product meets the requirement or not. A typical example of a non-testable requirement is, "The product shall be easy to use." A version of this requirement that is testable is, "The target user shall be able to perform basic functions (detail what functions) within 30 minutes after operating the product for the first time."

COMPATIBLE: Each requirement should not conflict with other requirements.

CLEAR: All the members of the product teams shall have the same interpretation of what the requirement means.

PRIORITIZED: Each requirement should be accompanied by a level of priority. Not all requirements are priority 1.

MODIFIABLE: Each requirement should be able to be modified without unsettling the entire set of requirements.

Step 7: Define and Document Requirements

Level 2 Main Process: Images and Statements to Requirements

Continued from Step 5: Conduct the Post-Visit Meeting

CONT'D

18

PERFORM THE AFFINITY PROCESS

This step may take up to three hours, perhaps more. See: Affinity Diagram.

Exceptions:

- Images always take priority over other statements; that is, they are always higher in the Affinity Diagram.
- The Affinity Diagram will yield that images. These are actual customer quotes that express their feeling about a need or describe a problem.

19

EXAMINE THE AFFINITY DIAGRAM

- At this point, your Affinity Diagram should have about five to seven key categories, or headers.
- These headers should be images.
- Below each image should be statements, requirements and, finally, features and other items.
- You should also have between 25 and 50 yellow stickies in total. If you do not, then you need to do additional consolidation. For example, combine several yellow stickies into a single statement that encompasses the overall voice of the customer.

See Also **Affinity Diagrams**

Main Header

The customer quote — the Image — that represents the highest level of the expression of need or description of a problem.

Key Categories

These are additional images or voices that reflect a close affinity to the main header.

Supporting Statements

Voices, requirements, features and other messages from the customers.

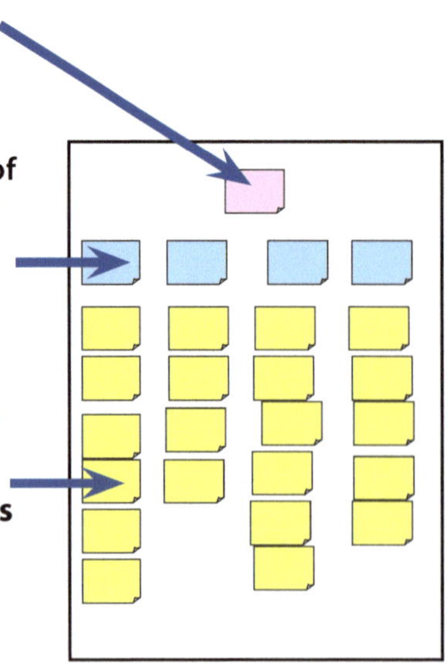

Important

Separate Images and Statements

This process can be non-linear, meaning that in some cases it will not follow the prescribed sequence. Notice that there is much dialog within your team; consequently there will be iterations and rewinding. Allow it to happen; it is important in order to discover valuable information. The relationship between the steps in this process is well-established; the dialog may be, and in many cases will be, non-linear.

CONTINUED

Step 7: Define and Document Requirements

Level 2 Main Process Cont'd.

Cont'd

20

Select the Images

Choose the yellow stickies that best represent the topic of the column. Although you already have "headers," you still need to examine the column to select the one quote from the customer that best represents the topic of the column. If no appropriate yellow sticky is found, leave the original header in place. It is always best to have an actual quote from a customer as the header of the column. If you do not have one, it is probably because you still need to make progress in your interviewing skills.

Remember that there is a difference between an image and a statement. Hierarchically, an image is higher as it contains more emotional energy — i.e., more of a feeling than the statement. Images should go higher in your Affinity Diagram and be associated with statements that relate to the same customer needs. Together they will form the base for defining good requirements.

Where You're At

At this point you and your team have organized all the relevant yellow stickies into categories, or headers. You did this through the use of the Affinity Diagram process.

The headers represent major clusters of critical information from your customer, categorized by affinity. Each header is an actual customer quote, which represents the entire cluster or you have created a header that represents the cluster.

21

Study the Affinity Diagram

You goal is to now develop a set of requirements.

You and your team should be familiar with section on Good Requirements in this guidebook and also the section on the Market Requirements Document (MRD).

Important

There will be several sets of requirements. Each major category in your Affinity Diagram will generate a set of requirements. You need to start a list of requirements using yellow stickies.

22

Develop the Requirements

Key points:

- This steps starts the process to develop the requirements for the voice of the customer.
- Create an open dialog to clearly understand the voices, images and other information in the Affinity Diagram.

See Also: Good Requirements

See Also: Market Requirements Documents (MRD)

Cont'd: Step 7: Define and Document Requirements

PHASE III

Innovate

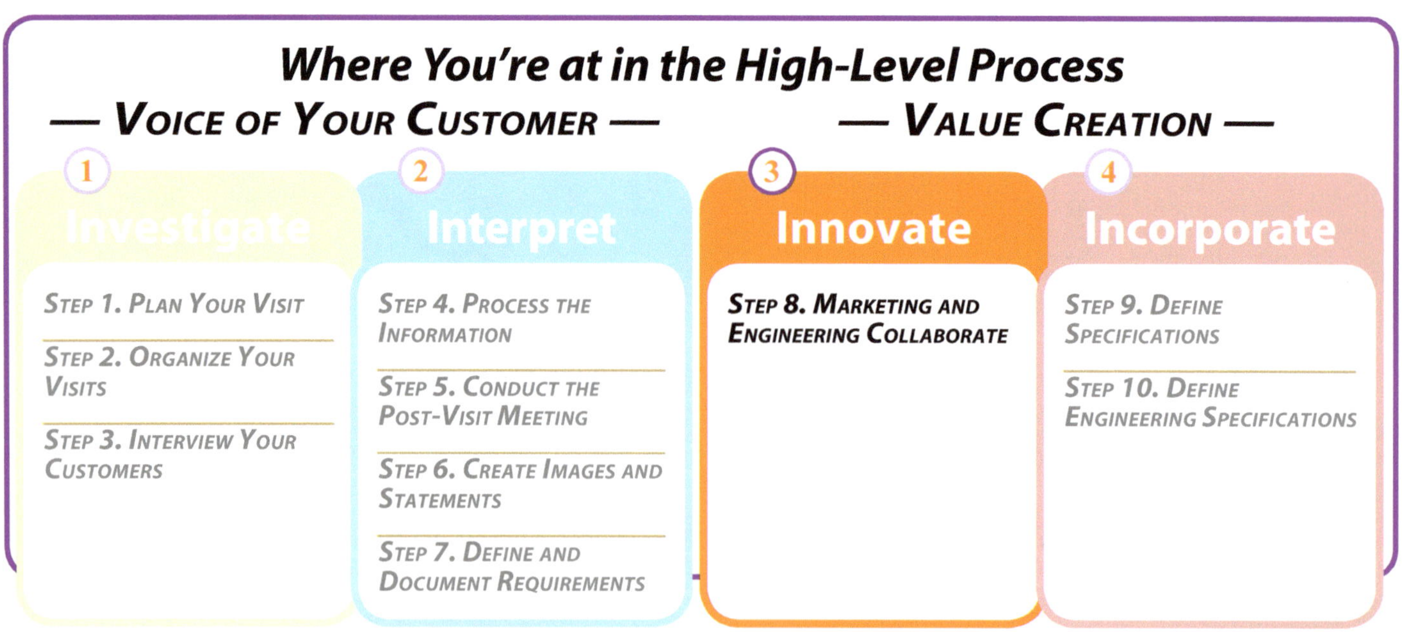

Step 8. Marketing and Engineering Collaborate

Create Something Better

Overview

Innovation is the ability to bring new value to your customers. Notice that we do not limit innovation to technology; rather, we suggest that you look at the entire set of experiences that your customers go through in solving their problems.

In technology companies the focus is generally on technical features; that is, the "banner specs," which indeed are very important — but not at the expense of other areas of innovation.

You are striving to differentiate your product and at the same time solve your customers' problems. We suggest that the more competitive your environment, the more you should think of innovation in terms of the entire customer experience.

The ability to innovate comes from clearly understanding customer requirements, and the ability of marketing and engineering to collaborate to address your customers' challenges.

Once your team understands customer requirements, set about addressing each requirement in an organized and methodical fashion. This generally involves meetings, which can last several hours and employ specific approaches — for example, breakthrough thinking skills and problem-solving skills.

At this point you have concluded the Post Visit Meeting (PVM) and you and your team have extracted a set of customer requirements. This step concludes the second "I" in our model: Interpret. Now you are ready to start the third phase: Innovate.

The good news, if you followed the process, is that members of your R&D organization have been active participants in the process — developing the Discussion Guide, going on customer interviews, turning voices into requirements, etc. — which enables them to better understand the customer requirements.

Having a clear, documented and prioritized set of customer requirements is essential to the innovate. Frankly, you cannot innovate without a set of customer requirements because if your R&D organization does not understand the problems that your customers are trying to solve then any innovation would be off target.

Directed Innovation

When you have a clear, prioritized and documented set of customer requirements, your R&D organization can direct their innovation energy to real customer needs, creating value for the customer.

Note that at this point you do not yet have features, but this is good! You do have customer requirements, which allow your engineering organization to solve real problems without any pre-conceived ideas of features.

The Next Steps

Update the Market Requirements Document (MRD) Using the MRD Process

It is essential that you now finish developing the best possible MRD. It's the way that you communicate requirements to your R&D organization, and it enables them to truly create innovative solutions for your customers (following this introduction there is a full section on the MRD, the process and suggestions for improvement)

Brainstorm Solutions for Each Requirement

At this point, you have done nothing more than transform customer inputs into a set of requirements. You have not yet discussed how these requirements could be achieved. This is the next step. Each requirement can have several solutions. It is at this level that creativity is needed. Many breakthrough products happen right at that level. Brainstorming is a good way for a team to generate possible solutions. Let's use our "Laptop" example:

- Requirement:

Step 8. Marketing and Engineering Collaborate

- A typical business woman shall be able to carry the laptop all day without fatigue.

- Possible Solutions for the Requirement:
  - The laptop and its accessories weigh no more than three pounds. (This has been found in other studies to be the limit for women.)
  - The laptop comes with a rolling carrying case.
  - The laptop is carried in a backpack.
  - The customer can ask somebody else to carry it for her.
  - The laptop itself can be rolled.

The last idea demonstrates that brainstorming always brings apparently stupid solutions, but they should not be disregarded at this stage.

Each solution is then discussed. And each of them brings a different approach to the problem. There might be opportunity to combine some of them for a more powerful final solution. In the case above, the three-pound solution is the preferred choice. But the team estimated that the technology was not ready for this kind of performance within the budget they had. So, they decided to have it as a long-term goal and to go with a combination of lightweight and a rolling case.

Express Solutions as Features

In order to design the product, engineers need to know **WHAT** they will design. So, the next step is to express the solution as a set of features. Continuing to build on the "Laptop" example, the features associated with the solution could then be:

- Lightweight.
- Attractive carrying companion.
- Easy to carry or roll.

Establish Specifications for Each Feature

We are nearly done. The last step is to assign metrics for each feature. Metrics are what Engineering will design against, what Manufacturing will test and what Marketing will list in the product data sheet — although not all specifications are tested or listed, some of them being purely internal. For example, the set of features above can be specs as follows:

- Lightweight: maximum 4.5 pounds.
- Attractive carrying companion. 90 percent of test-market subjects rate the carrying companion as attractive or very attractive and would definitely want to own one.
- Easy to carry or roll. 90 percent of test-market subjects rate the combination laptop plus carrying companion as easy or very easy to carry, 90 percent rate it as easy to roll, and 90 percent rate it as easy to go from carry mode to roll mode.

Step 8. Marketing and Engineering Collaborate

The Market Requirements Document (MRD)

Introduction

In most product development organizations, the marketing organization communicates customer requirements to the R&D team through a document generally called the MRD. In many cases, the document doesn't serve its purpose, which is to effectively communicate the prioritized customer requirements to the R&D team so that they can develop innovative solutions. In addition, this document often creates frustration, confusion and internal battles because the engineering team does not find what it needs. That leads to products that are late, poorly executed and fail more frequently than they should. One study associated 74 percent of failed products with deficient or missing market studies. In this section, we will discuss the problem, its causes and make four specific recommendations for changes that will build an effective bridge between Marketing and Engineering groups, create an environment better suited to respond to market and customer needs and increase the chances of success for your new products.

Definition

The Purpose of the MRD

To effectively establish a common understanding between Marketing and R&D of a prioritized set of customer requirements.

Definition

The Outcome of the MRD

To enable the R&D organization to create customer value through innovative products.

Current Misunderstandings

Too often the term MRD is incorrectly interpreted because of its name. Let's start with **MARKETING**: Marketing does not have requirements; customers do.

The second misunderstanding is about the word **REQUIREMENT** and what should be in this document. Many MRDs today do not include actual requirements, but rather a list of features generated by Marketing. These features are seldom justified in terms of why they are needed. They are often included for competitive reasons rather than to allow the customers to better address their critical needs.

Finally, effectively communicating customer requirements to R&D takes more than a single **DOCUMENT**. In reality, it should be treated as a succession of events; that is, a process. We advise you adopt the Customer Requirements Process (CRP). We further stress the importance of Marketing and Engineering (R&D) collaborate from the start.

Why the Misunderstanding?

The traditional MRD is a direct result of the "silo mentality" that prevails in some organizations. The MRD is the formal way of communicating between the Marketing silo and the Engineering silo. In addition, R&D often does indeed request a list of features to do its work. In these silo organizations, you often hear engineers say, "Tell me what to do, and I'll do it." This attitude should not be acceptable! The new trend is for Engineering (R&D) to fully partner with Inbound Marketing to obtain, process and prioritize customer requirements.

This state of affairs comes from the fact that the Product Manager or Marketing Manager responsible to develop the MRD is often performing the dual role of Inbound and Outbound marketing. Obviously, the tactical nature of Outbound Marketing always takes precedence, But this leaves little time to develop a useful MRD. Furthermore, both types of activity require vastly different sets of skills and abilities.

Least likely is the case where requirements come from an actual analysis of what customers want to achieve. Consider that if your MRD is simply a list of features, you have stifled Engineering's ability to innovate. The role of Engineering is to understand the customer problems and then design innovative solutions that create value; thus, when Marketing merely delivers a set of features it defeats that purpose for important new products.

Yet another reason for inadequate MRDs is that some marketing people do not clearly understand what a requirement is, nor the intent of the MRD. They see it as the way to convey a list of features for a new product without understanding the context or the problem facing the customer. Few companies have mastered the art and science of this process; it is difficult and takes both tenacity and a robust framework.

Step 8. Marketing and Engineering Collaborate

Contrast Between In-Bound and Out-Bound Marketing Functions

Product Planning / Inbound Marketing	Demand Creation / Outbound Marketing
Analytical by nature.	Tactical by nature.
High business acumen.	Excellent support person.
Ability to suspend the selling behavior.	Understanding of the selling process.
Profound connection with markets and customers.	Able to connect with sales and distribution channels.
Excellent listener.	Able to constantly think demand creation.
Able to connect with R&D.	Versed in marketing communication.
Ability to connect and influence company executives.	Good at managing multiple tasks.

This diagram is intended to show a few of the main differences in skills and abilities between In-Bound Product Planning Marketing and Out-Bound Demand Creation Marketing. The above is not a job description, but rather a contrast between the aptitude, knowledge and skills of two discrete functions within a marketing organization.

Note: The terms "in-bound marketing" and "out-bound marketing" have come to mean the generation of Web traffic — please be aware of possible confusion.

MRD Maladies

Most MRDs fail to effectively communicate customer requirements to R&D. Here are some of common causes:

The Cryptic MRD

The lack of useful information is why a cryptic MRD provides insufficient clarity for the engineering organization to develop a solution (i.e., create value). Vague statements such as "make it easy to use" or "increase resolution" are still found in too many MRDs. As a result, Engineering is at a loss as to how to proceed. Worse yet, it becomes the source of friction and conflict between Engineering and Marketing.

The Feature-Rich MRD

An MRD with a long list of features, but without the corresponding customer requirements, priorities and rationale, will severely reduce R&D's ability to provide innovative solutions. These features are often vaguely defined — e.g., "beat the competition" — without any rationale as to why they are needed by the customer.

This lack of valuable information drives the organization to develop an ill-adapted or general purpose product — one that fails to solve a few, well-defined customer problems because it tries (and fails) to solve too many problems. It often becomes the source of friction and conflict between Engineering and Marketing because all features are deemed priority one! The Feature-Rich MRD is the sign of a product- or sales-driven organization, but not of a market-driven one.

> **Important** If your MRD is simply a list of features — you have stifled the ability of R&D to innovate.

The Bureaucratic MRD

Some companies have forms and templates for their MRDs. While this can be very useful, it also opens the risk of MRDs being treated as routine or boilerplate information. They simply fail to get the in-depth research into customer requirements, but instead offer too much filler content.

The Confusing MRD

This occurs when an MRD is a very large document with too much information that is not relevant to customer requirements — it's full of text and information that have no relevance to the key topic of addressing critical customer needs. This extraneous information is a distraction and obscures the real issue, which is a clear and prioritized description of customer requirements.

The One-Sided MRD

This takes place when a marketing professional creates an MRD without the collaboration of the R&D organization, then he is surprised that it is rejected and conflict ensues. The creation of an MRD is a process where Marketing and R&D collaborate from start to finish; for example, representatives from both organizations conduct customer interviews, and together they extract requirements, and finally they collaborate in the creation of value.

The Current Situation

Marketing needs to reassess the way customer requirements are communicated and get away from the idea of a single document. Modern product development is too complex to depend on a single document. We suggest looking at various ways to help Engineering understand the requirements. Rich tools are available, such as videos of use cases, photographs of the major steps in the customer's process and in-depth interviews performed jointly by Engineering and Marketing. Other tools that facilitate the clear description of customer requirements include personas, scenarios, storyboards, process maps- and desired outcomes.

Requirements

1. Articulate Real Needs

Requirements articulate what the product must have or must do to solve problems or satisfy needs that your customers have expressed. Alternatively, it could also be expressed by what customers must be able to do or have as a result (outcome) of using the product. But — and that is where many marketing people fail — requirements should not specify how the product will do what it must do. That is why features and specs are not requirements per se. Let's take a simple example: A business traveler wants to use a laptop during a five-hour flight. Most people would say that the battery must last at least five hours, but that would not be a valid requirement by our definition. "Five-hour battery life" is one solution, but there are others. The correct, although incomplete, requirement is: **"The user shall be able to work on the laptop for five hours during air travel."**

2. Must be Verifiable

The second important thing to remember about requirements is that they must be verifiable. Otherwise, the product team will never know if the requirement is met. In the case above, the requirement is not 100 percent verifiable. What applications does the customer use during a long flight? Playing videos, for example, will require more power, and the battery will not last as long as with a simple business application. Therefore the conditions of use must be specified as closely as possible so that the requirement is verifiable. Thus our example could change to: **"The user shall be able to view movies on the laptop for five hours during air travel."**

3. One Reveals More

The third thing to note is that a single need expressed by a customer will generally dictate several requirements, some of them obvious and some not so obvious. In the previous example, the laptop and its owner do not land in the plane seat by magic. They must first pass security and the x-ray machine. That calls for some additional requirements. On the plane the user may be working with confidential company data; thus, another requirement. Furthermore, a requirement may need several features to be fulfilled; said another way, to satisfy a single requirement you may need an aggregate of several features.

Step 8. Marketing and Engineering Collaborate

There are other qualities that good requirements must possess. However if you only remember these three, you'll be way ahead of your competitors and on the road to great products — products that users will feel have been designed solely for them and that create a strong customer attachment.

Four Recommendations for Improvement

1. Use the Customer Requirements Process (CRP)

Turn the articulation of customer requirements into a process and away from a single document. Using a Customer Requirements Process (CRP) results in both Marketing and Engineering signing off on documented requirements in a way that works for the entire team. The CRP starts very early in the definition of the new product, where Marketing and R&D partner from the start. Together they identify customers to interview and set about performing the research. Then they proceed to use modern methods to describe customer requirements, including use cases, personas, scenarios, storyboards and process maps. And they agree on a clear priority.

Initial investigation and detailed VOC capture take place in the Investigate phase. Conversion to requirements and development of the Customer Requirements Document (CRD) follows in the Interpret phase. In order to keep the team focused on the customer requirements it's essential that discussion about features and solutions be delayed until the Innovate phase. But this topic has traditionally been part of the MRD. So, how do we handle documentation of features and market descriptions? The answer is to have a Product Requirements Document (PRD) that documents solutions and features, separate from the market-related information of the MRD.

Breaking up the traditional single document will go a long way toward making the product team market-driven because the first gate (approval of the CRD) will solely consist of reviewing customer requirements. The CRD becomes the document that is used as input for the process to create the PRD. The PRD articulates how the product will do what the CRD says it should do.

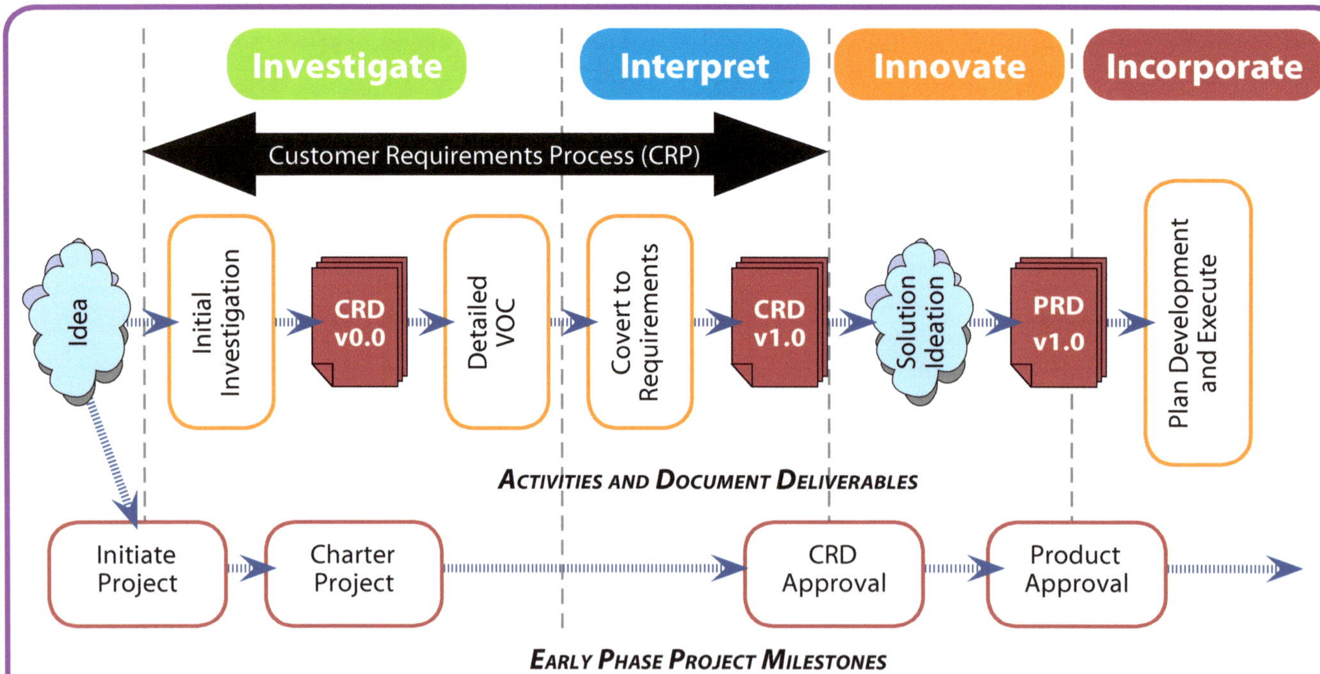

This diagram shows the suggested Customer Requirements Process (CRP) incorporated into a traditional initiation, or "fuzzy-front-end" New Product Development Process (NPD). Notice the Customer Requirements Document (CRD), which becomes part of the process.

Step 8. Marketing and Engineering Collaborate

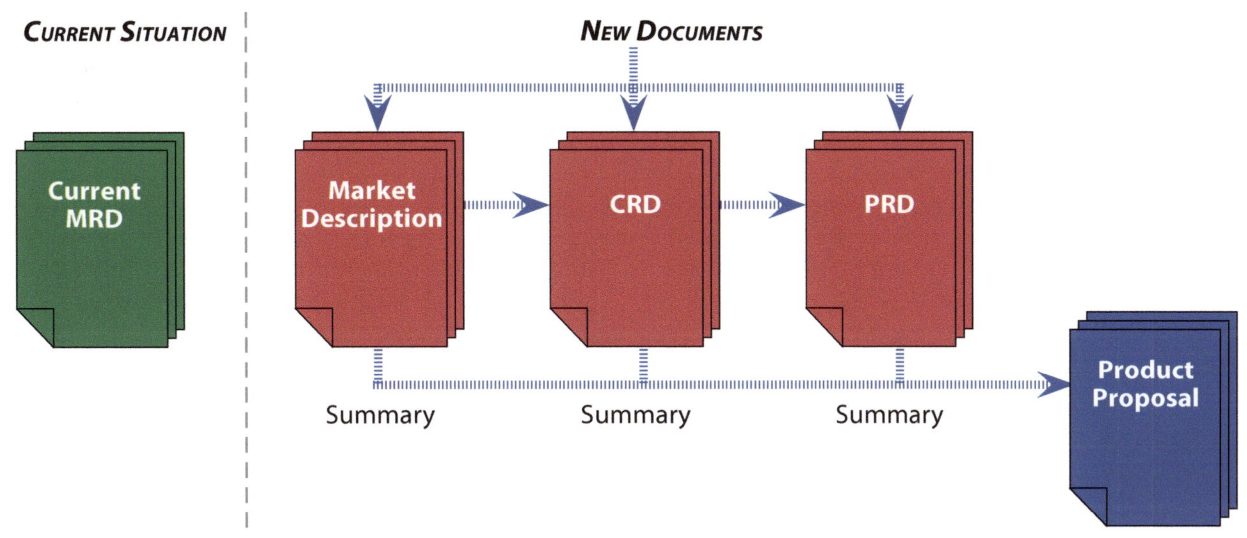

This diagram shows our suggested method for replacing the current MRD with three separate and distinct documents. This approach will ensure that the customer requirements are fully described and prioritized in their own document; thus, enabling clarity, ease of communication and the right level of attention.

Note: Regardless of the number of documents, the critical intent is to have all the information about customer requirements in a single and discrete focal point.

2. Forget About Features

As mentioned, the MRD often includes features. But a list of features stifles R&D's ability to discover richer solutions — it constrains innovation. Having the separate CRD and PRD keeps the team focused on customer requirements in the early stages.

3. Make Prioritizing a Priority

Having a list of clearly articulated customer requirements minus a list of features is a good start. However, if these are not prioritized and associated with target market segments, they will be useless. Prioritizing the requirements can achieve two things:

- Provide clarity to the team when trade-offs have to happen (typically time versus cost) during the development program.
- If the requirements are prioritized and associated with the relevant market segments, it can help in identifying business impact if features which support specific requirements are dropped during the development process.

4. Collaborate with R&D and Engineering from the Start

We recommend that Marketing and R&D partner very early in the development project and jointly discover and prioritize customer requirements. This should be done under the leadership of the marketing person. Resist the urge to start designing the product until the requirements are clearly understood. The job description of R&D must include the responsibility to understand customer requirements and their priorities.

The days of the lone marketer going out with his notebook to talk to the customer and coming back delivering his message to the R&D

The MRD is no longer a single document, but a process to describe, prioritize and effectively communicate customer requirements to R&D.

The MRD does not belong to Marketing — it is developed in full collaboration with R&D.

organization are gone. A market-driven organization demands a strong partnership between Marketing and Engineering (and Manufacturing when appropriate). Therefore, demand that engineers team up with Marketing to interview customers and capture voices.

We often hear some senior managers making the objection, "We don't have the time because of cost and time-to-market pressure." This is not valid! Similar false assumptions made strong companies believe that good quality costs more money. No! Good requirements and time-to-market are not contradictory. Good requirements shorten time-to-market. Organizations that have indeed adopted such a process report an actual shortening of time-to-market. Having clear requirements at the beginning leads to a clearer product definition, which in turn leads to faster implementation and fewer do-overs in the development phase. On the other hand, consider that many of the sources of delay have to do with poor or absent requirements, lack of prioritization, lack of understanding of the problems and outcomes that your customers expect as a result of using your product.

Conclusion

Poor or incomplete requirements is the number one reason why new products fail. In light of this, it's incomprehensible that so few organizations give the quality of requirements proper attention. There are two reasons this happens. The first occurs if the process to properly capture requirements has only recently been implemented, and Marketing and Engineering have not mastered it yet. This can be fixed by institutionalizing the new CRP process into the actual New Product Development Process (NPD).

The second reason is more pernicious. It's based on the false assumptions held by many managers that capturing requirements takes too much time, which will lengthen time-to-market and that, in any case, they already know what the market needs.

Developing a prioritized set of customer requirements where features are relegated to later stages in the development process and by having Marketing and Engineering collaborate creates clarity at the initiation stage. You will have a firm foundation on which to develop products that:

- Customers love to own and use.
- Improve business results.
- Reduce the waste of developing capabilities that your customers don't value and won't pay for.
- Accelerate time-to-market due to the alignment between Marketing and R&D and the development of only the capabilities your customers need.

Ten Questions a MRD Needs to Address

The authors have chosen not to provide a template for the creation of an MRD because it is much too important a document, and templates tend to automate an activity. Rather, we have opted to provide a set of critical questions that you can use to test the viability of your MRD.

Answer the following questions in the most succinct, specific and convincing way. It should take no more than a page and a half. If you can convince your audience in a page and a half, the rest is just supporting information.

1. What are the market, the market segments, and the targeted market segment(s)?
2. Why? Characterize the market thought a profile of its size, rate of growth, amount of growth, the current market share your products have, etc.
3. Who are the customers? Identify the economic buyer, technical buyer, user, influencer, customer's customer, etc. Describe key personas. Articulate the decision process to purchase your type of products. You can use the template on Decision Making Tree in the Sample Selection of this guidebook.

4. What is the Customer's Reason to Act (CRTA)? Define the customer problem that needs to be fixed or addressed that will cause him to act and purchase your product.
5. What are the Customer Value Drivers (CVDs)? Identify the criteria customers would use to make a purchasing decision? Look for CVDs in the five dimensions of value: technical, operational, economic, psychological (or emotional) and network.
6. How will you create value for the customer? What is the concept of the offering you are considering?
7. Why are you going to win? How will you create competitive advantage?
8. What is your proposed value proposition? The answer to this question also is a clear answer to the CRTA.
9. How will you make money or capture the value you provide?
10. What could make you fail? How will you manage these risks?

The above list should not be used as a template or a questionnaire. Rather, it should help you formulate a message to R&D to enable them to innovation and create value for the customer.

A Full Palette of Tools to Describe Requirements

Introduction

Communicating customer requirements to the R&D and the rest of your team can be daunting — at times it may seem almost impossible. Customers do not buy products or features; they buy solutions and benefits. Most of all, they buy differentiation; that is, the demonstrated ability of your offering to deliver greater benefits and value than that of your competitors.

All of the above can only start with understanding customer requirements, then implementing solutions, which then become your offering. Note that features and specifications are simply a means to an end.

Fortunately, there are many tools and methods available to effectively articulate and communicate requirements. It all starts with a profound understanding of your customer — a superficial understanding will not yield valuable requirements.

A typical customer requirements would read something like: "The user shall be able to work on the laptop for five hours during a flight." Note that there are no features or specifications in the statement, that comes much later.

In addition to the methodology to articulate a requirement statement there are additional methods to effectively describe requirements, here are a few examples:

Personas

The profile of a fictitious individual for the purpose of understanding the behavior, personality, experience and other personal traits of a typical individual in a market segment or a function. A Persona is a description of the education, training, experience, preferences, skills, abilities and other personal traits that help understand the user of your product at the personal level. For example, a Persona might describe the typical business traveler that uses a laptop during extended business trips.

Use Cases

A formal methodology to describe the specific action, thoughts, and steps that a particular customer uses to perform an action or her duties. Use cases describe the interaction between a user and a particular product or a between the user and a feature of a particular product. For example, a use case might be the description of the actions performed by a traveler when using a laptop during an extended flight.

Process Maps

The documented, sequential steps performed to achieve a result. Process maps are similar to Use Cases but the focus is on the accurate description of the sequential steps with minor focus on the person performing the steps. For example, mapping the process to set up and initiate a laptop in an airplane seat.

Scenarios

Synthetic description of an event. A scenario could be the description of what a customer does to perform part of his job. It is also a description of a particular event under a specific set of conditions that involve a customer and his reaction to it. For example, a scenario might be describing what happens to the user of a laptop on an extended flight during heavy turbulence.

Storyboards

Storyboards are the sequential visualization of a customer's activities. The methodology had its starts in the theatrical and film industries as a way to visualize the sequence of scenes in a film, opera or theater production. Widely used in software development, it is finding its way into hardware and other product development arenas. The methodology helps understand the process, interdependence and relationship of all the activities of a customer under a specific situation. For example, storyboard what a user does to get the laptop from his luggage, set it up and start working.

Time and Motion Studies

A technique started by Frederick Winslow Taylor, used to map, time and describe a task or a process with focus on the individual performing the task. The resulting data is analyzed to find opportunities for improvement and create solutions which might improve performance. For example, doing time-and-motion study on a person performing a set of tests on a device in order to create a better solution.

Semantics

The study of how humans interpret words and language. It is the difference that may exist between the meaning of a word or phrase as meant by the "sender" and the interpretation of that same word or phrase as interpreted by the "receiver". For example, the phrase "easy to use" may have simple meaning or may imply a highly technical one; it can also be misinterpret by the receiver, which in this case would be you, during a customer interview.

Heuristics

In the area of gathering of customer requirements, Heuristics is the ability to closely observe a customer performing a task or function in order to discover the thought processes, to learn the attitude and motivation of the customer and how customers solve problems. The observer then develops a set of Heuristics that describes the rules to be used in developing a new product. Heuristics are commonly used to determine the man-machine interaction, human interfaces and other areas related to ease of use.

Neurolinguistics

In short, the study of the neural framework of the human brain. Practically speaking Neurolinguistics is a multi-discipline area of study of how the brain processes information, and how this information becomes language and communication. For example, the study on how customers interpret a GUI or a set of instructions.

Conclusion

There are still more tools. Ethnographic studies, demographics, contextual analysis — there is no shortage of methodologies. While we do not recommend that you adopt all these tools, you should definitely build your own "tool box" appropriate to your situation.

There are indeed many tools, methods and templates to help you describe customer requirements with clarity. It all starts with your commitment to an in-depth understanding of your customers and their requirements.

Step 8. Marketing and Engineering Collaborate

Level 2 Main Process: From Requirements to Features and Attributes

Continued from Step 6: Break Voice into Images and Statements

WHERE YOU'RE AT

You now have a relatively short list of customers, perhaps 7 to 10. This list emerged from your Post Visit Meeting, and it's the result of the Affinity Process you used to create categories. For each category, create one or two requirements. This list is not yet prioritized

See Also: **Karnaugh Maps** — Prioritizing Items from a Section of Many

See Also: **Market Requirements Documents (MRD)**

CONT'D

23

PRIORITIZE THE REQUIREMENTS

Obviously, not all requirements are first priority. Your understanding of the market plus the series of interviews will provide the criteria for prioritization. We recommend using Karnaugh Maps to prioritize.

This is a critical step, as it will define the value you intend to provide to your customers. It will set the agenda for your R&D. And it will determine the margin you will obtain from the new product.

Take your time, involve your team and perform the proper analysis before moving to the next step. You may even have to call some customers back to get additional insights.

24

REVIEW YOUR MRD

You may have already started the MRD process, but now, it's time to update it — get it ready to include the prioritized set of customer requirements.

You and R&D must be in alignment as to the purpose and format of the MRD.

25

FROM REQUIREMENTS TO FEATURES

There are two paths:

1. **BRAINSTORM SOLUTIONS**: Not yet features, solutions are strategic approaches or concepts. We encourage you to think of solutions first because customers buy solutions to their problems and challenges.
2. **JUMP TO FEATURES:** This path is less desirable. The risk is that you would focus too much on a feature and forget the solution.

Important!

Moving from Requirements to Solutions to Features demands dialog and brainstorming. Don't short-change the process — this is where innovation and competitive advantages are developed. Don't relent — the process gets easier.

Methods such as Breakthrough Thinking Skills, Problem Solving, Root Cause Analysis, and others are very helpful.

See Also: **Ladder of Abstraction**

See Also: **Imaging Glossary**

CONTINUED

Step 8. Marketing and Engineering Collaborate

Level 2 Main Process Cont'd.

CONT'D

26 ...TO ATTRIBUTES

YOUR CHOICE

A This is the desirable path, from requirements to features to attributes.

B It is possible to go directly from requirements to attributes when the requirement does not call for innovative solutions to be found.

27

FROM REQUIREMENTS TO FEATURES

At this point, you and your team develop a set of requirements to match the voices (statements and images) generated from the Affinity Diagram category clusters and prioritization.

After naming your requirements, which may take several hours of dialog and brainstorming, proceed to prioritize the requirements. Use a formal method to prioritize. To achieve consensus and speed you must develop a set of criteria to use in prioritizing.

At times it may seem chaotic, but fear not! It's fine.

Definition

Attribute

A description (articulation) of an actual trait of the product, what a feature is or how the feature is implemented.

See Also

Critical Tools

Methods for prioritization

28

FROM FEATURES TO ATTRIBUTES

Attributes get you closer to an engineering specification.

- "Portable" becomes an attribute by saying, "It will not exceed 20 cm X 10 cm X 5 cm."
- "Lightweight" becomes an attribute by saying, "It will weigh 600 gr. or less."
- "Ease to use" becomes an attribute by saying, " The menu should not be more than two layers deep, no more than four colors will be used in the GUI, and Helvetica will be the only font allowed." Notice that to make the attribute real in this case, a set of features is required.

29

PRIORITIZE

This is a good time to step back and review your priorities. You may need to re-prioritize requirements, features, attributes, etc.

CONTINUED

Step 8. Marketing and Engineering Collaborate

Level 2 Main Process Cont'd.

Where You're At

At this point, you should have a documented and prioritized list of requirements, as well as a similar list of features and attributes.

In our experience, this will not yet be a final version of a single product. But do celebrate if you got one! Most likely you have several concepts or configurations, particularly if the requirements demand considerable innovation, such as when creating a brand new product or platform.

You may need to test the concepts or configurations with your customers. This requires additional customer visits and/or more market research. This is a critical point: you and your team must make the decision to test or not. We suggest that most of the time you will need to test. Therefore, don't shy away from the responsibility of testing further with customers.

30

From Features to Specifications

After you and your team have decided whether to test your concepts or configurations, proceed to define the features in technical terms.

Here's where engineering does the heavy lifting. Naturally, it will require much interaction, as trade-offs will need to be made. While engineering is in the lead, marketing must collaborate closely to ensure the integrity of the voice of your customer. Refrain from "throwing" documents between the two departments. Be actively collaborative!

Resist the temptation to move to features too early in the process — ensure that requirements are properly defined.

STOP

Step 8. Marketing and Engineering Collaborate

Imaging Glossary

Term	Definition
Attribute	A description (articulation) of an actual trait of the product, what a feature is or how the feature is implemented.
Benefit	The positive outcome that an offering provides to a customer. Widely used by salespeople, who typically want to transform the features and solutions given to them by the factory into benefits that can resonate with their customers. Bankable benefits are rooted in images expressed by the customer. For example, start with the image, "I carry this laptop all day from customer to customer. When the day is over, my shoulder hurts. I wish I had a laptop I can carry as conveniently as a book." Develop a solution that allows you to advertise "Thanks to its innovative design and material, you can now bring your laptop with you all day without fatigue — you won't even notice it at the end of the day." If the marketing job of transforming voices into specifications has been well done, the linkage should be immediately visible. When it is that vivid, the customer experiences elation at discovering the feature.
Feature	A description of the functionality or quality that will address or solve a requirement. Not to be confused with the benefit, which is the result that the customer will achieve from using the feature.
Image	A vivid mental picture that describes the problem, challenge or issue confronting a customer. Preferably, it is a verbatim phrase, sentence or narrative from your customers.
Requirement	Simple statements, usually containing the verb "shall," that describes the functionality (what the user or the product shall be able to do) or the quality (what the product shall have) that solves or addresses a need — for example, "The user shall be able to carry her laptop all day without fatigue." A requirement expresses what the offering should do — but not how it will be done — to address the need expressed in the image or, more generally, what the customer needs to do. If a customer says, "I need four-hour battery life," is it a requirement? Maybe from the customer's perspective, but not according to our definition. "I need four-hour battery life" is a specification (four-hour life) for a feature (internal battery), itself part of a solution (battery-operated) for a requirement and an image that have not been expressed. The chain is incomplete. The real requirement might justify (or not) the solution/feature/specification expressed by the customer. For example, "four-hour battery life" is not the only solution if a customer wants to work during a four-hour flight; in that case, the real requirement should be expressed as "The laptop shall be capable of operating for four hours on a plane."
Specification	The description of the attribute in terms that an engineering organization can design. For example: Weight, 5 pounds; size, 8x11 inches.
Value	The customer's subjective perception of what he or she wants to happen in a specific situation — and as the result of using your product.
Voice	A verbatim statement from the customer or a compendium of such statements that represents the customer reality.
Whole Product	The minimum set of products, services and intangibles necessary to ensure that the targeted customer will achieve his or her Compelling Reason to Act (CRTA). Whole products are responsible for the customer experience.

Incorporate

PHASE

Where You're at in the High-Level Process

— Voice of Your Customer —		— Value Creation —	
1 Investigate	2 Interpret	3 Innovate	4 Incorporate
Step 1. Plan Your Visit Step 2. Organize Your Visits Step 3. Interview Your Customers	Step 4. Process the Information Step 5. Conduct the Post-Visit Meeting Step 6. Create Images and Statements Step 7. Define and Document Requirements	Step 8. Marketing and Engineering Collaborate	Step 9. Define Specifications Step 10. Define Engineering Specifications

The Last Steps

NEW TOPIC

Work Smarter

"Clean Execution" is another name for this final step. Going from customer input to a Requirements Document is no easy task. Once the Innovate phase has taken place, the need to firmly add the new features into the product becomes imperative. The input from your customers must be unequivocally woven into your project. Generally this is known as engineering specifications or design constraints — it's Phase IV: Incorporate.

From Features to Engineering Specifications

The challenge of the Incorporate phase is to protect the integrity of the Voice of the Customer. Dialog wins the day!

Generally, engineering owns this process, and it is documented in your New Product Development (NPD). But engineering and marketing must collaborate in order to make the difficult trade-offs that inevitably arise at this stage; for example, technology may not allow a feature to be implemented, but a clever innovation may solve the problem. Formal meetings are required where the issues are discussed objectively. Follow up is required too — create formal documentation of the agreement, trade-offs, key assumptions, etc.

A feature may not be worded in sufficient detail for engineers to proceed with design and development. A process to translate the feature into engineering specs is needed so that the Voice of the Customer is not lost. From experience, we have found that MRDs (Marketing Requirement Documents) are insufficient for engineering; thus, a dialog must ensue. The Voice of the Customer must win.

Essential

"The challenge of the Incorporate phase is to protect the integrity of the Voice of the Customer."

Engineering Change Notices (ECNs)

As the development progresses, an ECN gives engineering a process to manage changes. We discourage the use of ECNs to add features to a product because it disrupts the development process and creates unpleasant trade-offs. The major features of a new product should be defined at the start of the process. While markets do change, these changes do not happen overnight. Thus, maximum effort by marketing to define the major features of the new product at the front end is imperative.

In the rare occasion an ECN is needed to add a feature, the same process must be used as in the start; that is, from "voice" to "feature" to "engineering specification." It must have the same discipline. It must have the same attention and focus.

High-Level Process

Major Steps

Step 9. Define Specifications

- Prioritize the innovations from the previous phase.
- Prioritize the attributes of your new product.
- Prioritize the features of your new product.

Step 10. Define Engineering Specifications

- Engineering does the heavy lifting in this phase, in collaboration with marketing, to ensure the preservation of the Voice of the Customer.
- Incorporate into the design documentation. (Engineering specifications.)
- Monitor progress and adjust to reality.

Methods and Tools Used

General Methods

- Requirements formulation
- QFD (Quality Function Deployment)
- Documentation and document control
- Advanced communication skills

Specific Methods

- Conjoint Analysis
- Kano Model
- QFD (Quality Function Deployment)
- Requirement management
- Rapid prototyping

Continuous Improvement

Your Company

Understanding your customers takes tenacity, commitment and a lot of work. It also requires that your organization implement the methods, processes, tools and templates needed to support your effort.

We highly recommend that you gradually implement all the needed changes. Given the current nature of business — with all the urgency and chaos — it is best to lay the foundation and then set about making gradual, but consistent improvements.

There are many approaches to continuous improvement; your company may already have a structure in place. If so, take advantage of it. If not, you might consult any of the many books on continuous improvement available in the marketplace.

Further, the need to understand that focus on customers must become part of the culture — the DNA — of your organization. For this to come about, you will need the clear support of your executives and managers. They hold the key to cultural changes. If they do not value your goals, you will have a difficult, if not impossible, task of changing the culture by yourself.

Your Career

It behooves you to embrace the need to understand your customers. This perspective will help you throughout your professional career. Always think about your customer first. Always continue to learn about better ways to understand customers. And make a commitment to champion your customers regardless of what your current organization or colleagues say.

Critical Tools

Cascading a New Tool to the Organization

Introduction

There is a difference between attending a seminar or workshop and causing your organization to embrace any or all of the tools you learned. Said another way, just because you and a number of your peers attend a seminar, that's no guarantee that what you learned will become common practice in your organization.

Deploying a tool takes additional work. It is a process, and requires planning, tenacity and leadership. Consider that you are not just enabling the adoption of a new tool or method, but actually changing old habits. In short, you will cause disruption and a break from the comfort zone currently in existence in your organization.

The Business Case

The return on investment (ROI) from any seminar or workshop only occurs after the class and when visible and measurable improvement in productivity is achieved. Some have the erroneous notion that attending a seminar is the objective; instead, attending is a means to and end. The real objective is to take what was learned and apply it to improve productivity.

Expect Resistance to Change

However, just because you and a number of your peers attend a seminar or workshop, that's no guarantee that what you learned will become common practice in your organization. Deploying a tool takes additional work. It's a process that requires planning, tenacity and leadership. Consider that you are not just enabling the adoption of a new tool or method, but actually changing old habits. In short, you will cause disruption and a break from the comfort zone currently in existence in your organization.

Expect resistance to change in overt and passive ways when you suggest the adoption of a new tool or method. You will hear things like "It won't work here," "We tried that, and it didn't work," "Our situation is different," and many similar statements. There are numerous resources about driving change in an organization to help you.

The Hesitaters

The initial push-back may actually be the expression of other feelings:

- **SEEKING ADDITIONAL INFORMATION:** Some in your team resist as a way to obtain additional information, as they feel fearful to deploy something they do not understand. It's the best way they know for doing fact-finding.
- **SEEKING REASSURANCE:** They feel unsafe; for example, they want to know that they will not get fired if the tool does not work the first time out.
- **TESTING:** They are simply trying to determine if this is for real and if you have a strong conviction that the tool will work.
- **GENUINE CONCERN:** Some may actually feel that the tool will not work or is not applicable.

The way to deal with all the above feelings is to first be aware that they are common, and then validate the actual source of the concern by providing additional information, reassurance, etc.

> Important
>
> Much that has been written about driving change in an organization. When searching for books, seminars, articles and other resources use the terms "Managing Change" and "Resistance to Change."
>
> Also, your human resources department may be able to help or offer additional resources.

The Reactionaries

Some individuals are simply opposed to change, reactionaries who want to preserve the status quo. Occasionally you will run into someone who will refuse to change, refuse your new tool and steadfastly remain opposed. It is rare that you are able to change these individuals, and we don't have the time here to delve deeply into this

Deployment of a New Tool

topic. We suggest that you do not become discouraged; rather, build a constituency of those who do want to change, and use it to overcome the resistance.

Food for Thought

The reluctance to change is a part of the human condition; you are not a bad person if you resist change. Your strategy to overcome resistance must take into account that it is a normal reaction. Further, the process to overcome resistance is all about people. Do not rely on templates, software programs, process maps, etc. Rather, seek to understand the situation, and then tap your leadership skills to drive the change.

Here are a few things to keep in mind:

- First, set about understanding the current situation of those who need to adopt the new tool or method. Talk to them as a group and, more importantly, individually. Make an effort to listen with the intent of understanding.
- Be sure to consider WIFM (pronounced "wiffem"), which stands for "What's in it for me?" We react positively when we understand our personal benefits. You must make the case for embracing the change such that every person on the team sees his or her benefits as part of the deal.
- Stop selling and start communicating! Our excitement for the new tool gets the best of us, and we start pushing the idea to the point where it becomes annoying. Instead, remain open to suggestions. Some will try to help you succeed by providing ideas for improvement — for example, a better way to use a template or a better way to deploy the tool. Be open to ideas, and engage those who provide them; do not reject them either directly or passively.
- Speak their language. Adapt your message to the audience. Speak in a way that is familiar to them. This will improve communication and avoid the use of new words.
- Start small. The intent of the first use of your new tool or method is to train and to develop the organization; thus, do not select a major project. Rather, find a simple project where the use of your new tool will shine and the risk of failure is very small.
- Finally, your tenacity and leadership will carry the day. Don't be disappointed with setbacks. In the end you're doing the right thing for your team and your company.

The Critical Questions

All of us need a certain amount of information before we commit to a tool, method or practice. It is not fair to expect people to jump at the opportunity if they don't have sufficient context, especially at the personal level.

Generally, we all want the following questions answered:

- Why are we doing this?

  (What is the relevance to my team, our company or our organization?)
- What would happen to us if we don't do it?

  (What are the consequences to the company, organization or team?)
- Why are you asking me to do this?

  (Tell me why I should care. What's in for me?)
- What support can I expect from you?

  (Can I rely on you? What is your commitment to me?)
- What specifically do you expect me to do?

  (Tell me specifically how this impacts my job. What specific changes must I make?)

The above information must be communicated in the language and context of the listener. Avoid platitudes and irrelevant information as they tend to confuse and add uncertainty.

Level 3 Detailed Process: Deploying a New Tool

1

Choose the Tool

It's best to start small. Deploy a single tool to reduce the challenge and risk of failure.

Define clear criteria; for example:

- Ease of learning.
- Amount of leverage the tool will provide.

You can select any criteria, but you need to be consistent in your evaluation and prioritization.

Start recruiting supporters and allies who react positively to the new tool.

Important

Use the Planning Document in this section to ensure that you are well prepared for the deployment.

2

Identify Your Core Team

You will need a small group of your peers to help you deploy the tool.

Team members should be willing and able to spend the needed amount of time helping you with the task. Sometimes it may take up to four hours per week.

Make sure the team will be in place for at least three calendar months. It might take longer.

Important

Remember to secure the visible support of your organization's executives. Without it, you risk failure.

Meet with your executives, and show them the benefits and improvements that your new tool will bring to the organization.

If possible, secure a sponsor or a mentor — one of the executives who is willing to assist you in the deployment.

Seek support from others who are willing to help deploy.

3

Secure Support from Your Executives

It is prudent to identify one of the executives of your organization to sponsor your efforts.

Sponsors, also called Mentors, will provide moral support, help remove obstacles and add a measure of stature to your efforts.

The best sponsor is one who, like you, has a passion for the tool you are leading.

Important

As you select the pilot for your deployment, also ensure that you have training, templates and other needed support to help the team succeed.

Continued

Deployment of a New Tool

Cont'd

4

Identify the Pilot

This is the team, organization or group of employees that will apply the tool to a real project.

The results of this pilot will provide valuable information when deploying the tool to the entire organization.

Remember to obtain approval from the appropriate managers and supervisors.

Important

Equip the Pilot

It is critical that you provide coaching, encouragement and support to the pilot team. There is the risk of push-back and disappointment if the tool does not work as expected

It would be best if you were part of the pilot team. That way, you will be there throughout the process.

5

Conduct the Pilot

Ensure that you provide a high level of support to the organization such that the deployment of the new tool does not end in failure.

Provide all the training, tools, encouragement and other forms of support throughout the deployment of your pilot.

Important

What if the Pilot Fails?

Even the best of efforts may results in failure.

The first thing to do is to debrief and identify the root cause of the failure — without this step you cannot recover.

Avoid blaming, guilt or punishment as these things add no value to the recovery process.

Once you have the root cause and lessons learned, modify the process and try a new pilot.

6

Debrief the Pilot

The topic of the debrief is to identify what worked and what did not so you can make appropriate improvements for the full deployment.

This debrief is done with your Core Team.

7

Proceed with Full Deployment

Deployment of a New Tool

Planning Checklist

Deploying a tool is indeed a project. Consequently, ensure that you have a deployment plan, a schedule, budget, etc. Working with your Core Team, or with the peers who have agreed to help, you can develop a tactical plan that will ensure a successful deployment.

The list below is intended to help you formulate your own plan. You may need to add your own items to the list. Also note that the list below is in no particular sequence. Finally, the list is not rigid. Use your judgment and do what makes sense within your organization.

Planning Checklist

- ❑ Schedule for the deployment defined, documented and approved.
- ❑ Training materials for employees developed and documented.
- ❑ Training materials produced — hard and soft copies — available for use.
- ❑ Training materials placed in the intranet or web page.
- ❑ Trainers of the new tool identified and informed of their duties.
- ❑ Training of employees scheduled and employees notified.
- ❑ Training of managers and supervisors scheduled.
- ❑ Feedback mechanism in place.
- ❑ Metrics to measure the progress and eventual success of the deployment identified and ready to use.
- ❑ Coaching and encouragement for managers, employees and stakeholders defined and ready — including those who will do the coaching.
- ❑ Support available; for example, facilitators, coaches, tech support, assistants, etc.
- ❑ Obstacle removal in place.
- ❑ Templates, checklists and other tools ready and available.
- ❑ Information on our web page done.
- ❑ Processes that will be impacted by the deployment identified and corrected.
- ❑ Changes to our quality manual completed.
- ❑ Manuals, guides and other reference materials that will be needed by the users ready and available.
- ❑ Executives briefed and ready to fully support the deployment.
- ❑ Human resources (HR) informed.
- ❑ Training department informed.

Deployment of a New Tool

Planning Document

This template may help you secure assistance from your executives. It is also a good point to start the process of deployment. It may even become a standard template for deployment of future tools and methodologies. Modify this template to meet your specific objectives within your organization.

1. Tool or methodology to be deployed.
2. Objective of the tool deployment.
   - What will be the outcome once the tool has been successfully deployed?
   - Why are you deploying this tool?
3. Who is the expert on this tool?
   - Who is the internal person who knows the tool and is willing to provide guidance to others?
4. What documentation is available?
   - Models.
   - Software.
   - Books.
5. Are there any external resources available?
   - Consultants.
   - Instructors.
   - Facilitators.
6. What are the risks involved in the deployment?
7. How will we pilot the tool?
   - Which team, organization or group will be the first to apply the tool to actual work.
8. What will be our communication plan?
   - With executives of the organization.
   - With employees.
   - With the pilot team.
   - With other stakeholders.
9. What will we measure to assure us that the deployment has been successful?

Deployment of a New Tool

Tips for Successful Deployment

There are a myriad of things you must consider before you deploy a new tool in your organization.

Brainstorm with your team to find innovative ways to ensure the success. Your corporate culture, industry and reporting structure will help you determine the best way to embed the new tool.

Key Questions

- How can we make the new tool a habit in our organization?
- How do we turn the novelty of the tool into the normal way we do things?

10 Tips for Successful Deployment
1. START SMALL — PICK A SIMPLE PROJECT FIRST *The intent at this stage is to ensure that your team knows how to use the new tool.*
2. CONSIDER THE CULTURE OF THE ORGANIZATION *If your organization has not used this tool before, you need to lead your team to understand and embrace it. Some will push back, and you need a plan to deal with some skepticism and denial.*
3. PROVIDE THE RIGHT TRAINING TO THE RIGHT PEOPLE AT THE RIGHT TIME *Your team, and even some managers, may need training. You must provide the skills necessary to succeed.*
4. ENSURE CROSS-FUNCTIONAL PARTICIPATION *Most tools provide maximum benefit when people from several functions participate. Moreover, most projects involve people from several areas of the organization, and they must be included.*
5. PUT IN PLACE A PROACTIVE FEEDBACK MECHANISM *Enable your team to provide you with input on the new tool.*
6. BE AVAILABLE TO REMOVE OBSTACLES *Inevitably, your team will run into obstacles when implementing a new tool for the first time. Put in place a mechanism that will enable you to remove these obstacles; otherwise, the team may reject the tool altogether.*
7. DEVELOP AN ESCALATION STRATEGY *Some of the issues and challenges that will emerge in deploying your new tool will need the attention of your managers and other executives. Plan an escalation process before things come to a boil.*
8. MAKE INFORMATION ABUNDANTLY AVAILABLE *When first deploying a tool, your team, managers and executives will need information. Make certain that information is abundant and conveniently available to all.*
9. MAKE COACHING AVAILABLE *Your team members will need one-on-one assistance. Even if you provide training, individuals will have questions and other barriers that you must help remove.*
10. CELEBRATE *Recognize progress — celebrate small and large victories. Everyone needs reassurance that progress is being made.*

Yellow Sticky Protocol

NEW TOPIC

Effective Group Participation

Benefits

The ability to ensure that everyone in a group participates is greatly enhanced through the use of Post-It Notes® (trademark of 3M), known in the vernacular as "yellow stickies." Another benefit of yellow stickies is that they provide a self-documenting process; rather than writing things on a flip chart, which is time-consuming, yellow stickies significantly accelerate the documentation. Finally, yellow stickies are portable; that is, one can move them, hand them over, rewrite them, etc.

Some individuals, particularly senior executives, are not fond of yellow stickies. This is due to a misunderstanding as to who uses them. But this is a serious tool that will produce better results, and more quickly. These people need to be reassured that, barring a technological breakthrough, yellow stickies are still a fine tool that everyone, including executives, can use.

The process favored in this guidebook relies heavily on the proper use of yellow stickies.

> Essential
>
> "The process favored in this guidebook relies heavily on the proper use of yellow stickies."

Establish the Protocol

To derive the benefits of using yellow stickies, the facilitator must establish a protocol — an articulated set of guidelines as to the way the information will be presented in each yellow sticky.

The protocol is a way to create a safe environment. For example, use the protocol to assuage members who may have reservations about their poor spelling, grammar or handwriting.

The protocol is a way to choreograph the process so it yields maximum benefits.

For the sake of productivity, let alone sanity, the protocol must be enforced. All team members must participate and follow the protocol. Every yellow sticky must conform to a standard to be useful.

There is flexibility, so you need not do it exactly as in the corresponding detailed process. The overall goal is to be able to track and process a large amount of information.

Your team must agree on the way the yellow stickies will be structured. Once you have reviewed and highlighted all of your notes, you proceed to transfer the information to the yellow stickies. You need a protocol as to the type of information and its location. This is critical to provide continuity, traceability and consistency

Plan Ahead

Schedule a three to five hour meeting for a team of up to 15 people in a large conference room. Prepare topics and objectives for the meeting as well as the introduction. Gather materials:

- Post-It Notes® (3M brand, model #655 3x5).
- Sharpie® (Sanford brand, fine point, black markers) for all attendees.
- Flip chart paper and easels — two are best.

Level 3 Detailed Process: Yellow Sticky Protocol

1

Prepare your Materials

- 3x5, 3M-brand, Post-It Notes® (#655) — One pad per participant. Consider that several of the participants will have a sizeable number of yellow stickies — in some cases as many as several hundred.
- Sanford-brand Sharpie® markers, fine point, black.
- Flip charts or a place to hang the yellow stickies once filled.

Important: You also will need the participation of everyone who made customer visits.

2

Make the Announcement in Front of the Group

Face the audience, make eye contact and say...

"We are going to use yellow stickies in this meeting.

There are three key reasons why this is important:

- It is a self-documenting process, so it saves time and allows for documentation after the meeting.
- It allows everyone to participate.
- We can move the yellow stickies around depending on our needs — no need to rewrite, just move them.

Important: Caution: Some individuals, particularly senior executives, are not fond of yellow stickies. This is due to a misunderstanding as to who uses them. But this is a serious tool that will produce better results, and more quickly. These people need to be reassured that yellow stickies are a fine tool that everyone can use.

3

Explain the Protocol to the Group

"In order to take full advantage of the tool, we must follow a simple set of rules..."

For example:

1. Don't worry about spelling or grammar; no one will check it. This goes for everyone in this room.
2. Don't write on the sticky side of the note.
3. Write landscape (horizontally) oriented.
4. Use your Sharpie® marker. This will force you to write larger than usual and more legibly, and it will enable others to read notes from a distance.
5. If you make a mistake or change your mind, simply toss it and start a new one.
6. Write only one item per yellow sticky. Only one question, one idea, one quote.

Important: Encourage everyone to write legibly. Ask them to print. Many of us do not have the best of penmanship. Explain to the participants that they are writing for others to read! It will save time and reduce confusion.

4

Start the Process

See Also: Affinity Diagrams

Yellow Sticky Protocol

Example: Yellow Sticky Protocol

The topic of conversation from your discussion guide.

This must match the topics contained in the discussion guide that was created for this set of customer interviews.

An agreed-to indication of the importance, relevance or some other priority. In this example, "H" indicates "High" passion or importance of the statement from the source, whereas "M" would indicate "Medium" and "L" would indicate "Low."

Your team can develop similar priorities, depending on the objectives of your visits. Having a priority set will save considerable time during the later stages of this process. Keep it simple; otherwise, your team members will be confused and misinterpret the code.

This area is reserved for the message from the interviewee — the voice of the customer.

It can be a quotation, a fact, a finding, information or any sentence that is relevant. This is extracted from the notes you took during the interview — you must be able to trace it to your notes.

The Protocol

Topic	H	Customer's title
	"Actual quote from the customer, extracted from the interview notes."	
Your name/initials		Page, Interview #

The upper right corner is reserved for the title or function of the person who was interviewed.

You need to develop codes to be used here to help you understand the context, particularly if your team interviewed customers from different functional groups.

The name of the customer is not relevant.

A simple code will enable you to trace the information to your notes, and thus to an actual interview.

It is difficult to remember who said what. During the processing and various meetings, you will want to refer to your notes to clarify the context of the statement.

This is simply a tracking mechanism. You should be able to track back to the actual source of the information.

Quote marks (" ") are reserved for actual customer quotes! This is a critical rule.

Quotes are always verbatim messages from the customer. Do not use quotes if the customer did not actually say it.

Example

Ease of use	H	Design engineer
	"More than three layers of menus make products difficult to use."	
Fred T		Page 1, 2nd interview

Yellow stickies are landscape (horizontally) oriented)

This example shows a super set of the Yellow Sticky Protocol. You need not use all of the items, just those that will enable you to track and code the critical information. When you have several hundred yellow stickies, you will rely heavily on the protocol to make sense of it all.

Affinity Diagrams

Organize Information and Reach Consensus

Often, teams have to collaborate to solve a problem, develop a strategy, group options, identify alternatives, brainstorm, discuss a set of issues, etc. Tap into the power of team collaboration by enabling everyone to participate.

Affinity Diagrams have been used in quality management for many years. They are a tried-and-true tool. Once you master the tool, you can use it anytime you have to organize or prioritize information.

Historical Background

Affinity Diagrams belong to a set of tools that are called "KJ Method" or "KJ" for short, the initials of the Japanese anthropologist, Kawakita Jiro, who introduced the approach in 1967. The approach has its roots in the values of Zen Buddhism.

Mr. Kawakita developed the method as a means of organizing diverse observations and qualitative information into useful, documented facts. He observed that humans are not linear thinkers; that is, ideas flow in a non-linear fashion. Humans tend to jump around when brainstorming; consequently, trying to structure brainstorming sub-optimizes the creative flow.

Definition

Affinity

A natural liking for or inclination toward somebody or something or a feeling of identification with somebody or something.

Benefits

- Affinity Diagrams are simple and non-threatening, which creates an inviting environment for full participation.
- They are inclusive. Properly used, an Affinity Diagram allows everyone in a team to fully participate, regardless of rank.
- They ensure consensus in a group, as well as foster the unity of the team.
- They are a practical way of organizing seemingly unrelated facts.
- They are very conducive to the clear communication within the team.

Definition

Affinity Diagrams

A tool to organize information and reach consensus."

General Recommendations

1. Planning your session is essential to success. Please do not treat an Affinity Diagram process as a casual affair. Consult the process map in this document for additional information.
2. Select a room large enough to allow the team to move around. You may have to move furniture to create a clean space to work.
3. Affinity Diagrams work best when done by a group of about 10 people. Larger groups tend to fracture, meaning that some tend to marginalize themselves and not participate. Thus, groups larger than 10 require considerable facilitation and constant focus to ensure productivity.
4. Silence or talking? The general rules indicate the whole process should be done in silence to allow creativity to flow and avoid any one individual to influence the group. We believe talking can be desirable in certain circumstances. When you have a small group, say about three to four people, or when everyone is a peer and collaborative by nature, these are examples of when talking might help. Use your judgment, and decide based on the dynamics of your group.
5. Use the traditional rules of brainstorming — for example, do not allow criticism of ideas, encourage participation by all team members and generate the largest possible number of ideas, including duplicates and repetition.
6. Remember to follow-up. Many fine sessions end up delivering little or no value due to the lack of action after the session. Ensure before you start that you are committed to following up well after the session

Affinity Diagrams

is completed. Every session must have an action plan (roadmap) to ensure that you produce tangible results.

7. It's best to use flip chart paper. Other types of large-format paper such as butcher paper or newsprint rolls work too. Create a large space with paper — six to nine horizontal linear feet of paper is best. Remember that your team will be working in this area.
8. Avoid cryptic, one-word statements like "energy" or "easy." Encourage team members to be descriptive — for example, "There is not enough energy" or "Make it easy to use." Statements should have a noun and a verb to avoid ambiguity and enhance clarity.
9. The number of categories or columns should be kept to about five; more than that confuses the teams.
10. Create the feeding frenzy. It's fun and productive to encourage the team to generate as many ideas as possible. Set challenges — for example, ask everyone to generate five ideas. Do many rounds, create contests and give a prize to the person who generates the most ideas.

High-Level Process

1. **Start**
   Deliver instructions.
2. **Brainstorm**
   Participants create yellow stickies.
3. **Post**
   Place yellow stickies on the wall.
4. **Organize**
   Participants organize yellow stickies into categories.
5. **Prioritize**
   Participants prioritize categories.
6. **Develop Action**
   Assign responsibilities and develop a timeline.

Level 3 Detailed Process: Affinity Diagrams

1 Invite the Participants

Who should attend?

- All stakeholders in the process or project.
- Customers, if appropriate.
- Try to keep the group to 10 participants or fewer.

2 Prepare

- A large conference room
- Post-It Notes® (3M brand, model #655 3x5)
- Sharpie® (Sanford brand, fine point, black markers) for all attendees
- Flip chart paper and easels — two are best

See Also Yellow Sticky Protocol

3 Set Up the Topic

- State the topic in the form of a question. Ensure that everyone understands it. You may want to discuss it with the group.
- Reassure everyone that there are no wrong answers.
- Reassure everyone that grammar and spelling are not important.
- Explain how to use the yellow stickies — one per subject, use a Sharpie® pen, landscape (horizontally) oriented.
- No criticism.

4 Start the Process

- Give the question or topic, then ask everyone to write at least three yellow stickies and wait.
- Ask each person to read his yellow stickies as loudly as possible.
- Tell everyone to listen intently as others read their yellow stickies.
- Encourage participants to write additional yellow stickies as others present theirs.
- Collect all yellow stickies and place them on the flip chart or board at random.
- Continue until the group runs out of ideas or repetition sets in.

5 Scrubbing the Affinity Diagram

Clean up your Affinity Diagram once all the yellow stickies are written and placed on the board.

- Eliminate yellow stickies that are obviously irrelevant or trivial.
- Eliminate obvious duplications.
- Rewrite yellow stickies that are difficult to read.
- Rewrite yellow stickies that are not clear.

Important: For any and all changes you make when scrubbing the Affinity Diagram, make them with the agreement of the authors.

Continued

Affinity Diagrams

CONT'D

6

START THE AFFINITY PROCESS

Rally the group to move up around the flip charts with the yellow stickies.

Ask participants to observe how the various yellow stickies cluster or group into subjects.

Ask the participants to start moving the yellow stickies, using their own criteria to cluster them, regardless of how others feel about it.

Do this in silence! No one can talk during this step!

Stop when there is little or no movement of yellow stickies.

7

PICK 'N CHOOSE

Review each column or cluster that was created in the previous step, with participation by the group.

Ask the group to explain why the cluster or column exists. "Why are these yellow stickies together?"

Move the yellow stickies that do not belong in the column to a different one or a separate place on the flip chart.

Eliminate duplicates by choosing one that represents the topic and placing the others underneath.

For each column, assign a header that captures the topic. Write the header on a sticky notes, preferably a color other than yellow.

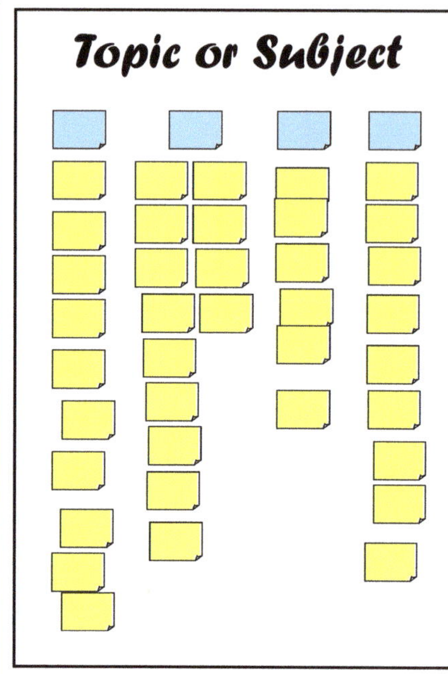

Yellow Sticky Approach

8

PRIORITIZE

Once the notes are prioritized, you can begin to assign responsibilities and action items as appropriate.

Also, develop a timeline. This is a critical step!

Save the flip charts for archival purposes or to be transcribed if necessary.

Debrief the group on the effectiveness of the tool and ways to improve the process.

See Also: Prioritizing Affinity Diagrams

See Also: Karnaugh Maps

Prioritizing Affinity Diagrams

The Process to Prioritize

Most often, Affinity Diagrams are used to prioritize a set of ideas, choices, things, etc. The criteria for prioritizing are determined by the initial objective that you set for the Affinity Diagram — for example, you might want to do an Affinity Diagram to assign responsibilities or action items. You might also want to pick a few from a selection of many. In short, use the same objective you set for the Affinity Diagram to select the method you will use to prioritize.

Group Consensus

Prioritizing is all about making decisions; keep this in mind as you work with your team. You want to reach consensus. 100 percent agreement is desirable, but not essential. Most of your prioritizing efforts will not have 100 percent consensus. That being said, prioritizing an Affinity Diagram is a team effort.

Select The Criteria

The key to prioritizing is the selection of the criteria used to make the decisions. Take your time, discuss with your team, and above all ensure absolute clarity. Criteria should have two or three elements; in a team environment and without computers it's very difficult to keep track of more than three criteria.

This process should be driven by the strategic importance of the project. For small, simple projects the criteria might be somewhat subjective. If the project is critical, then the criteria must be defensible — objective, verifiable and in full alignment with the strategic goals of your project. This is a qualitative exercise, so avoid criteria that are too complex.

Common Methods

1. ABOVE OR BELOW THE LINE

Draw a horizontal line at the midway, and ask the team to decide for each yellow sticky if it should be above or below the line. This is the quickest approach and one that most people find intuitive. You can prioritize the yellow stickies or the headers depending on your objectives.

2. 1 ROUND OF DOTS

This is a popular method because it's simple and familiar. Give everybody five to 10 self-stick colored dots to place on the yellow stickies based on the predetermined criteria. They can place their dots anywhere, as many dots as they wish per item. When done, simply count the dots to establish priority.

3. 2 ROUNDS OF DOTS

A variation on this theme is to do two rounds of dot placing with five dots for both rounds:

Round 1: Place five dots anywhere, with no guidelines.

Round 2: Place five dots, but only on two yellow stickies.

4. 3 ROUNDS OF DOTS

A more sophisticated method involves giving each team member 30 dots to be placed during three rounds:

Round 1: Place 10 dots anywhere, with no guidelines.

Round 2: Place 10 dots, but no more than two dots on any one yellow sticky.

Round 3: Place the remaining 10 dots, but only on two yellow stickies (five and five).

5. YOUR INVENTION

Invent your own system for prioritizing. As long as it's rational and leads to clear decisions, you'll be fine.

Prioritizing Affinity Diagrams

Level 3 Detailed Process: Prioritizing Affinity Diagrams

1

SELECT THE CRITERIA

The set of criteria is generally two to three elements that will determine priority. The criteria will drive the decision process.

Take your time, discuss with your team, and above all ensure absolute clarity.

The simplicity or complexity of your criteria depends on the strategic importance of the project. As a rule, keep it simple.

Once agreement has been reached, write the criteria on a flip chart to help the team stay consistent.

Important: This process should be driven by the strategic importance of the project. For small, simple projects the criteria might be somewhat subjective.

If the project is critical, then the criteria must be defensible — objective, verifiable and in full alignment with the strategic goals of your project.

2

START THE PROCESS TO PRIORITIZE

Q. Do we prioritize the headers, or each yellow sticky?

A. It depends on your objectives:

The preferred method is to prioritize the individual yellow stickies. Based on which categories the prioritized yellow stickies are in, you can infer the priority for the categories.

For example, if you use one of the Dots methods, count the number of dots in each category, and write it on the header. This will create an overall priority ranking.

Another approach is to ask the participants to place the dots on the actual headers — not the yellow stickies. This is faster, but it may miss an important message: Which yellow sticky is most important?

Use your judgment as to which voting method will serve you best.

Important: Prioritizing is a team effort. 100 percent agreement is desirable, but neither essential nor expected. The goal is consensus.

3

PRIORITIZE

Choose from one of these common methods or invent your own:

- **ABOVE OR BELOW THE LINE:** Draw a horizontal line, and ask the team to decide for each yellow sticky if it should be above or below the line.
- **1 ROUND OF DOTS:** Give everybody five to 10 self-stick colored dots to place on the yellow stickies based on the criteria. They can place their dots anywhere, as many dots as they wish per item. When done, simply count the dots to establish priority.
- **2 ROUNDS OF DOTS:** A variation on this theme is to do two rounds of dot placing with five dots for both rounds:
  - Round 1: Place five dots anywhere, with no guidelines.
  - Round 2: Place five dots, but only on two yellow stickies.
- **3 ROUNDS OF DOTS:** A more sophisticated method involves giving each team member 30 dots to be placed during three rounds:
  - Round 1: Place 10 dots anywhere, with no guidelines.
  - Round 2: Place 10 dots, but no more than two dots on any one yellow sticky.
  - Round 3: Place the remaining 10 dots, but only on two yellow stickies (five and five)

Prioritizing Affinity Diagrams

Example: Affinity Diagram

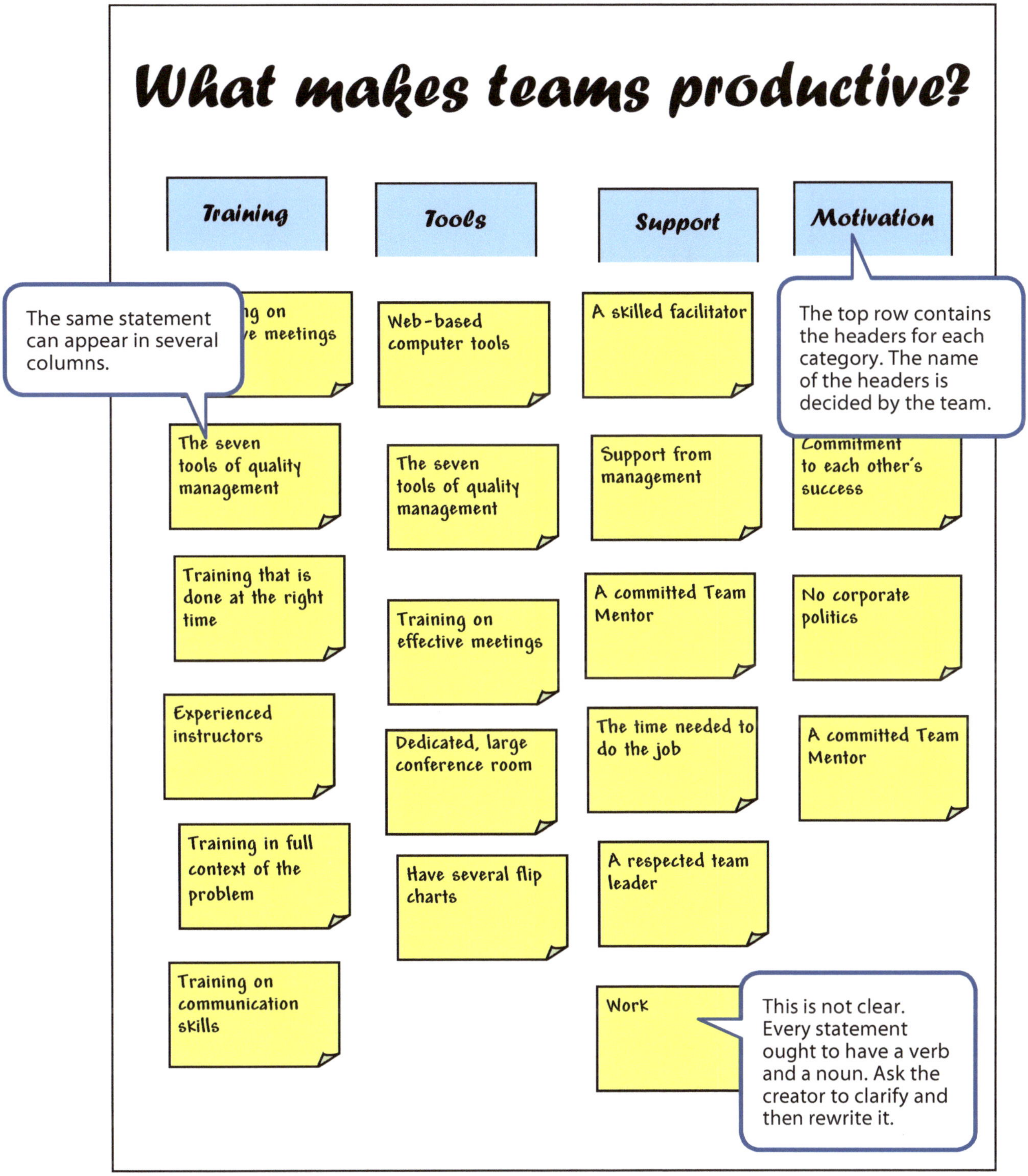

Karnaugh Maps

Prioritize Items from a Selection of Many

A Karnaugh Map is a participative 2x2 matrix used to make better and faster decisions. By varying the criteria and values, your team will be able to prioritize or select a few items from many.

Karnaugh Maps are a simplification of a tool called FMEA (Failure Model and Effect Analysis). They have their origins in mathematics, specifically as a way to represent a Boolean function. They are useful for generating the most optimal Boolean expression, and are related to Venn diagrams.

In general, Karnaugh Maps allow for two variables (criteria), called a pair-wise comparison, to be represented on the X and Y axes. You can also introduce a third variable by the color or size of the sticky note. We suggest you limit your Karnaugh Maps to three variables so as not to confuse the team.

Benefits

- Graphical and simple, they enable group participation.
- They engage everyone; thus, fostering innovation and breakthrough.
- They use two or three criteria, whereas other methods only use one.
- They are very flexible in their application to different criteria and values.
- They can be as rigorous as you need them to be — the results can be made statistically defensible.

See Also
Pair-Wise Comparison

Applications

Some specific applications of Karnaugh Maps might be to determine risk factors of a new deployment, select key features on a new product or determine the importance of tasks or projects.

For example, you can better allocate limited resources using the two criteria "urgent" and "important." While some things can be very important, they are not necessarily urgent, and vice versa. Thus, the ability of a team to separate all tasks based on the set of criteria will result in a better deployment of limited resources.

Use Karnaugh Maps to discover the Voice of the Customer. For example, the X-axis is "cost to develop" and the Y-axis is "importance to the customer." You can also introduce a third variable; for example, the color or size of the sticky note is "difficulty to develop."

Another example is risk selection. Take any project, and ask your team to identify 1) what is likely to go wrong and 2) what will be the negative impact on the project if it does. These two criteria will help the team focus on those things that are most likely to happen and would cause serious damage to the project if they did. Thus, rather than focusing on all the things that could go wrong, they focus on the vital few.

Another Way to Prioritize Affinity Diagrams

When prioritizing Affinity Diagrams, Karnaugh Maps provide more focus, and perhaps a bit more accuracy, than other methods — e.g., Above or Below the Line, Round of Dots. Karnaugh Maps allow for smoother interaction with your team, and they are a better way to communicate your priority or selection to your managers.

General Recommendations

- You should be familiar with the concepts of Affinity Diagrams.
- A session to develop a successful 2x2 matrix will take from three to four hours.
- The challenge is to select the right criteria and scales. The Y-axis and X-axis represent two criteria. The color or size of the stickies can represent a third. We recommend you start with just the two criteria.

The Scales

The scale is determined by the intent of your diagram. Generally, a scale of 0 to 10 on both axes is sufficient. When Karnaugh Maps are used for brainstorming, take a qualitative approach and choose a simple scale such as "0 to 5" or "small, medium, large."

When refining your selection or when the topic is of extreme importance, you may want to take a quantitative approach and choose a more accurate scale such as "0 to 100." We highly recommend that you choose such precise scales sparingly. In most cases, the additional accuracy is overkill.

Regardless of the scale, the team must agree on what each incremental level of the scale signifies.

More than Three Variables

Can you use more than three variables? Yes, you can. But it's important to show the results in three variables. You can accomplish this using a weighted average method. And you can determine a weighted average through the use of an Excel spreadsheet. In fact, using Excel, you can have as many variables as you wish.

Be careful! There is the tendency to over-complicate matters just "because you can do it." Let your objectives determine the number of variables. Also, we suggest that in most instances you will only need two, or at most three variables.

Example 1: Karnaugh Map (2x2 Matrix)

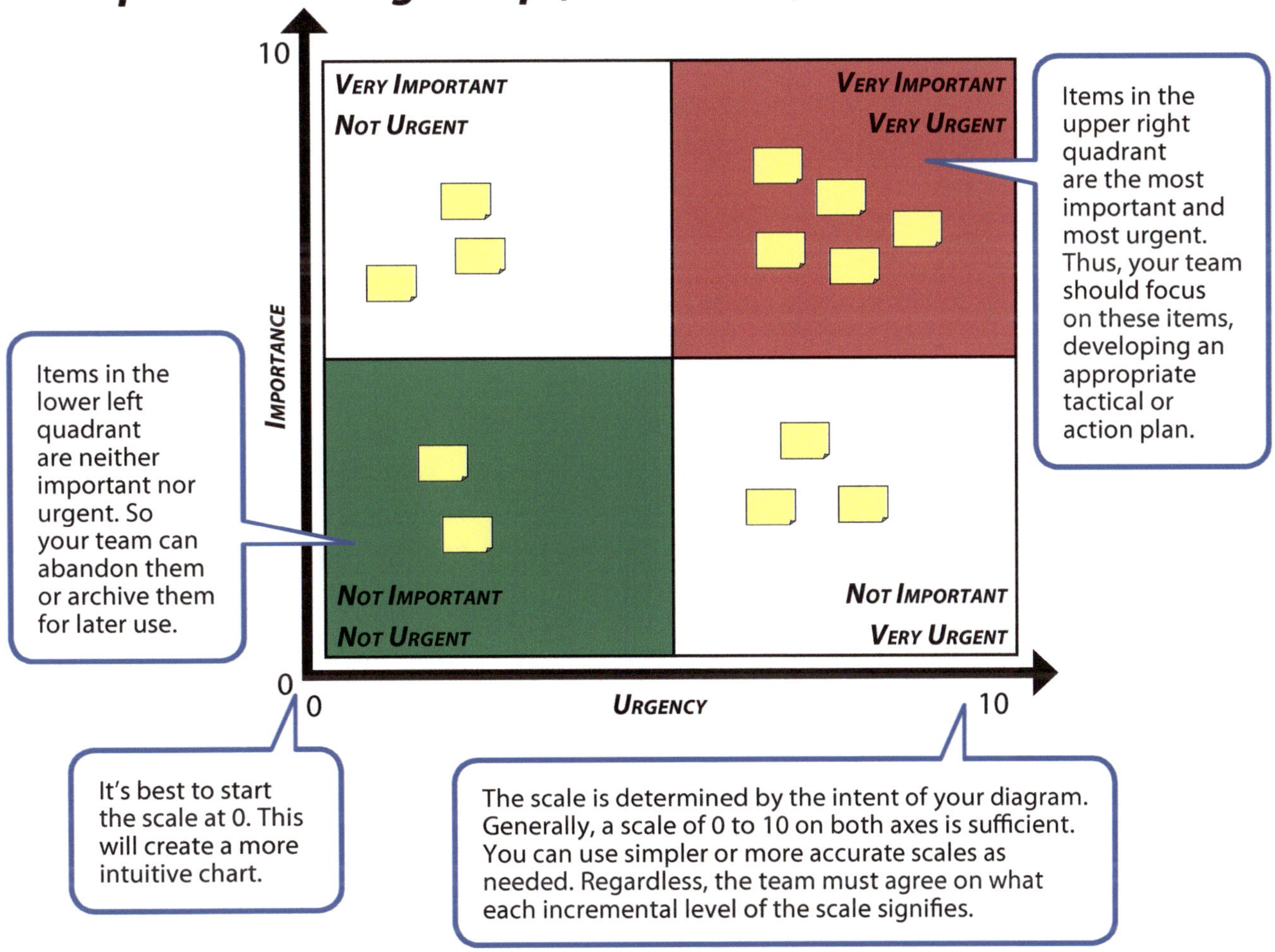

Karnaugh Maps

Example 2: Prioritizing Customer Requirements

Deliver the Value to the Customer Quickly

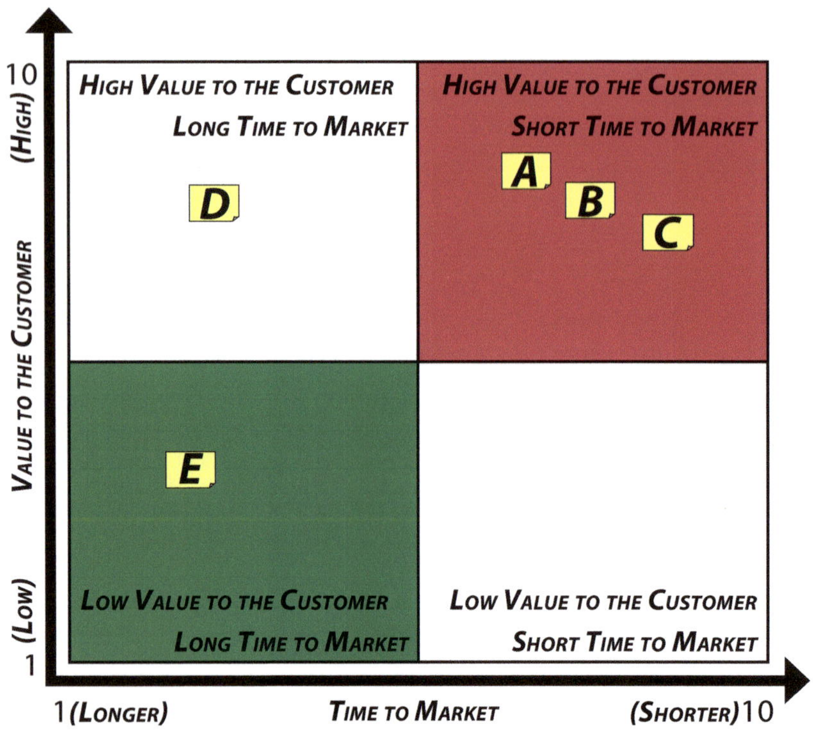

Here's an example of prioritizing based on pair-wise criteria: "Value to the Customer" and "Time to Market."

Items "A", "B" and" C" represent desirable candidates for high priority. Not only would these items deliver the highest value to the customer, but also they take the least amount of time to develop. In contrast, item "E" is not desirable at all.

Deliver the Value to the Customer in a New Way

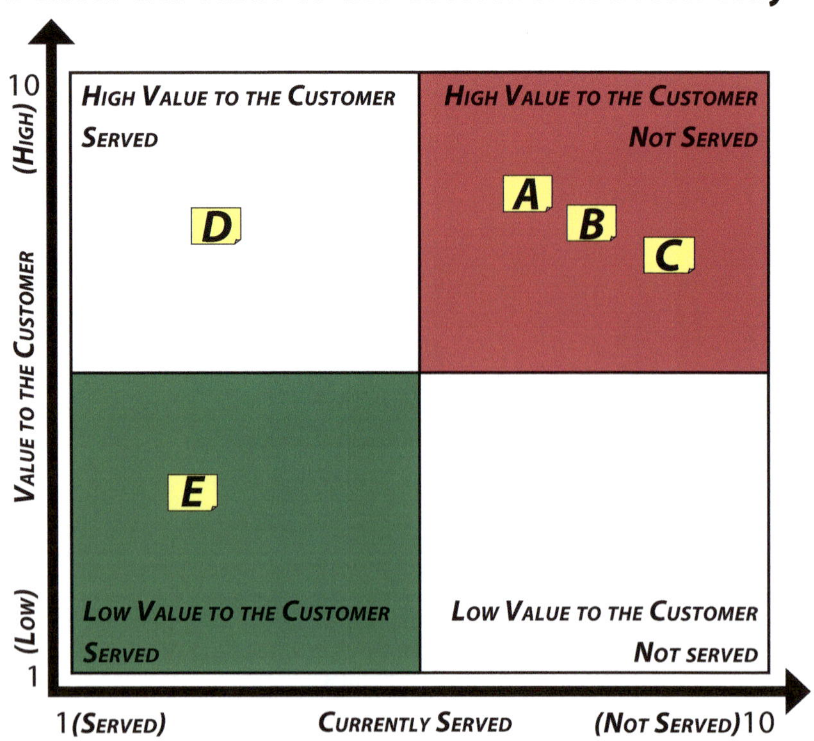

Here's an example of prioritizing based on pair-wise criteria: "Value to the Customer" and "Currently Served."

The variable "Currently Served" indicates how well the customer requirement is being addresses either by your product or by the products from your competitors.

Items "A", "B" and" C" represent desirable candidates for high priority. Not only would these items deliver the highest value to the customer, but also they are requirements that are not currently addressed. Again, item "E" is not desirable at all.

Pair-Wise Comparison

Pair-wise comparison generally refers to any process of comparing entities in pairs to judge which of each pair is preferred or has a greater amount of some quantitative property. Pair-wise comparisons are relational — unlike Cartesian X and Y coordinates, where one is the function of the other. The method of pair-wise comparison is used in the scientific study of preferences, attitudes, voting systems, social choice, public choice and multi-agent AI systems. In statistical analysis, it's used to determine probability.

Prominent psychometrician L. L. Thurstone first introduced a scientific approach to using pair-wise comparisons for measurement in 1927, which he referred to as the Law of Comparative Judgment. Thurstone linked this approach to psychophysical theory developed by Ernst Heinrich Weber and Gustav Fechner. In psychology literature, it is often referred to as Paired Comparison.

Application to Product Development

In product development, your team can use pair-wise comparison to express a priority between two mutually distinct alternatives in order to meet an objective.

Many pair-wise criteria are possible. We offer a short list of examples to help you and your team develop criteria that fit your needs exactly.

1. **CHARACTERIZING RISK:** The criteria of "Likelihood of Happening (Probability)" and "Severity if it Does Happen (Impact)" — often expressed as a P-I Factor — maps risks that are most likely to occur and will have the most severity if they do. You can use the results to manage risk.
2. **SELECTION OF OPTIONS:** For example, when selecting projects or courses, use "Attractiveness" and "Cost" to determine best value.
3. **MAKING CHOICES:** When making trade-offs, try "Ease of Use" and "Cost of Implementation." "Achievability" and "Benefits" are simple criteria you could develop to have real meaning — for example "Achievable within the Fiscal Year" and "Financially Beneficial."
4. **DEFINING PRIORITY:** Use "Urgent" and "Important" to prioritize the allocation of resources.
5. **DOING A GAP ANALYSIS:** Use "Current Level versus Desired Level" and "Difficulty in Achieving."
6. **DEFINING COMPLEXITY:** Use "Complexity" and "Difficulty in Achieving."
7. **DOING RESOURCE ALLOCATION:** Compare "People" or "Resources" on the Y-axis with "Priority of Action" on the X-axis. The priority can be defined in strategic terms or some type of urgency scale.
8. **DEFINING THE COMPETITION:** Map the need to defend a segment, abandon it or create a new segment. One example: "Level of Competition" and "Urgency to Defend." Another example: "Level of Competition" and "Critical Market Segment."
9. **ENSURING FIT:** Use "Fit with Business Objectives" and "Cost of Implementation."

Karnaugh Maps

Level 3 Detailed Process: Prioritizing a Karnaugh Map

1

Set Objectives for Your Karnaugh Map

For example:

- Determine risks and rewards.
- Prioritize from a choice of many.
- Select options based on certain criteria.
- Choose a course of action, given certain criteria.

Important

It's a good idea to prepare a blank 2x2 matrix in advance. Tape two or more sheets of flip chart paper together and use markers to draw the matrix. A white board is an alternative, but upon completion you need to carefully gather the yellow stickies.

Important

After a brainstorming session, you may end up with as many as 300 yellow stickies. Scrubbing is the step to eliminate duplicates, redundancies and any other non-value-added yellow stickies.

Your goal is to end up with approximately 30 or fewer yellow stickies for the last step. On the other hand, if you end up with many more, and you feel they are important, then proceed.

2

Process the Information

Once all the yellow stickies have been collected, start scrubbing the information to ensure that every yellow sticky communicates a clear, distinct idea.

This step can take 30 to 45 minutes.

If you have too many yellow stickies, then you need to affinitize (group) them into main categories.

Important

The selection of the two criteria for the X and Y axes is critical. To help determine the criteria, spend sufficient time with the team discussing "why" or "how" the final information will be used.

3

Define the Criteria and Scale

Select the pair-wise criteria for the X and Y axes. For example:

- Important vs. Urgent
- Expense vs. Ease of Implementation
- Risk vs. Rewards
- Important to Customers vs. Expense to Develop

Set your scale. For example:

- Small, medium and high
- 1 to 10 or 1 to 100
- A, B and C

Important

You can define a third criterion via the color or size of the stickies. Use this option sparingly. Furthermore, when first deploying the Karnaugh Map tool, we recommend you begin with just two criteria.

Continued

Cont'd

4

Plot the Information on the Matrix

Using your criteria, place each yellow sticky inside the 2x2 matrix. Do this in collaboration with your team.

You can use several methods of voting to assign value to each of the yellow stickies — for example, count hands, use dots, assign numerical values to later average. The more critical the topic, the more formal the vote should be.

Don't dispose of the plot, as you may need it for future reference.

5

Reach Your Conclusions

Placing all the yellow stickies does not complete the process!

The value of your 2x2 matrix is in the decisions that you make, the action-items that are generated and the follow through to act on these directives — simply put, the process is not over until something happens.

6

Develop Your Action Plan

Remember, it's not an action plan unless it clearly identifies "do what," "by whom" and "by when."

Document your action plan.

Follow up until all actions have been completed.

Process Mapping

NEW TOPIC

Steps to Success

A process is a **DOCUMENTED**, sequential list of **VALUE-ADDED** steps, the aggregate of which is used to deliver a product to a **CUSTOMER**.

- **DOCUMENTED:** A process must be documented; otherwise, it fails the most critical test. Furthermore, documenting the process has several advantages. It forces the dialog to be collaborative. It allows more than one person to study and use the process. Also, it's the only way to initiate formal improvements.
- **VALUE-ADDED:** Every step in the process must add value, as defined by the customer, or it is waste that must be removed.
- **CUSTOMER:** Every process must have a customer. For example, a customer of Marketing is Engineering. Thus, Marketing is the Process Owner, responsible for creating the product requirements, and Engineering is the Process Customer. Engineering has a set of needs, and it is the responsibility of Marketing to identify and satisfy these needs.

Create a Process Map

Creating of a process is not difficult. You can start by reviewing the many process maps shown in this guidebook. Collaborate with your peers, and define a process using the Yellow Sticky Protocol.

1. Call a meeting with people closely involved in the process you have chosen to document. For the process of producing a new brochure, for example, you would likely invite marketing staff, your advertising manager (marcom) and your advertising agency if you have one. These meetings generally last from two to three hours, sometimes longer.
2. Start the meeting by defining the process you wish to document — in this case, the creation of a new brochure. Also define the start and end of the process. Your team will help you.
3. Define the customer. In our example, the process customer happens to be your company's external customers. Be careful to differentiate between customer and stakeholder. In our example, the marketing manager is not a customer, but a stakeholder with a vested interest in making this process work. Furthermore, your general manager is also a stakeholder.
4. Define the output (result or product) of using the process. In our example, it may be "a printed brochure, ready for delivery to our customers." This critical step clearly defines the endpoint.
5. Brainstorm with your team to identify all the steps. Using yellow stickies, one for every step. At this point, you don't need to be concerned with value, sequence, duration, etc. of the steps.
6. Organize the steps sequentially, from process start to end. Add and refine steps as needed. Eliminate redundant or trivial steps. Take your time; this adds much clarity to your process map.
7. Study your map to ensure accuracy and clarity. Is the sequence of steps the way things really happen? Are there any steps that don't make sense? Are the steps described in a way that's easy to understand?
8. Test every step against the definition of the process. Pay particular attention to the value-added criteria. Ask yourself, is this step necessary? Does it really add value? This is your first opportunity to improve efficiency.
9. Finally, review your process map one last time, document it for permanent use and start using it.

Continuous Improvement

Once you begin using your documented process, the payoff begins. But the payoff continues as you make continues improvements to your process map. As you use it you will discover obvious ways to improve it — for example, redefine, reorder or eliminate steps. The secret is to constantly review your process map to identify opportunities for improvement.

Customer Panels

Interviewing Groups

Customer Panels is a generic term for customer contacts that involve more than a few customers at the same time — for example, when you meet with five customers from the same company. The Customer Panels model is a high-leverage activity, perhaps one of the most productive ways to obtain customer input.

Conducting a meeting with several customers is different than talking to a single person. The logistics are different; for example, you may require a conference room. The way you ask a question will change, and the way the question is answered will change.

The concept of Customer Panels is very flexible. It can be part of a regular set of meetings with customers, such as an ongoing User Group. It can be a group of users of your product. You can even meet with your competitor's customers.

Applications

Customer Panels shine when you need to get to the full story — why customers do what they do. Any time you seek in-depth information about the needs of your customers is a good time to do a Customer Panel. The panel provides a forum that allows for extended conversations with several customers at the same time. This interaction creates an atmosphere that unearths information not available from a one-on-one session.

- Use Customer Panels when Focus Groups fail to provide the depth of information you need. Many times, Focus Groups and similar tools do not help customers articulate the actionable information you need.
- Another reason to use Customer Panels is when you are after a solution and not a feature — for example, defining a brand-new product where there is no track record of a standard set of features.
- You can talk to several customers that represent a single function — for example, the end users of your product, or the managers who approve the purchase of your product.
- Use Customer Panels to take advantage of events where your customers gather, such as during a trade show or an industry conference.
- If you are traveling a great distance, Customer Panels maximize your investment.
- If you can only visit one customer site, conduct both one-on-one meetings and Customer Panels.

General Recommendations

- Customer Panels require advanced facilitation skills — for example, keeping a group on topic, ensuring that no one customer dominates the conversation, monitoring everyone's needs. Pick someone in your organization who has been trained in facilitation skills or hire an outside facilitator.
- Logistics are critical! Every part of the meeting must be carefully planned, from the recruiting of the participants to the venue and refreshments.
- Carefully train your team members who will participate in the meeting in areas such as communication skills and dealing with complaints.
- From the start, ensure that everyone, including your customers, knows the who, when, why and how of your panel.
- Your team must meet before you hold the panel to plan, understand the objectives and collaborate in developing a successful meeting. In addition, your team must meet after the meeting to properly digest the information and take action on what your customers said.

Customer Panels

Duration

The length of the meeting can vary a great deal. Plan for at least two hours, but half-day and full-day meetings are more common and preferable. More time allows customers to fully immerse themselves in providing you with valuable information.

Your objectives will impact the time frame. For example, if you have just a few topics to discuss, then the meeting might only take a couple of hours. On the other hand, if your objectives include usability testing, the meeting will take at least a half-day.

The availability of your customers also determines the time frame. You may need to be resourceful to get customers for a full-day meeting, but it is very achievable. Maybe you offer an incentive, such as a prize. Maybe you hold the meeting after work hours or on a weekend.

Focus Groups Vis a Vis Customer Panels

Focus groups are a tool of market research. They consist of a group of customers that have been recruited by a market research firm for the purpose of a group interview. Typically, a focus group meets in a conference room that has a one-way mirror to allow for observation.

While useful, focus groups have several limitations for defining new products. First, the dynamics of the group allow for one person to influence the rest. Second, there is generally not enough time for depth. Finally, the moderators are generally not versed in any one industry; thus, they are limited as to how deep they can follow a particular topic.

Focus groups are applicable for general exploration of a topic, perhaps at the beginning of a major development program. Use them to explore and generate ideas, and perhaps to create visions and tap the power of brainstorming.

As a real-world example, the authors combined elements of a focus group to create a Customer Panel. A market research firm was retained to recruit customers based on a profile provided by us. The firm also helped with the logistics. But the actual interaction with the customer was led by our engineers. It was very effective and a good example of synergy.

Level 3 Detailed Process: Customer Panels

1 Develop a Focus Statement

Don't avoid or postpone this step! The Focus Statement is critical to the success of a Customer Panel.

It should be developed in full collaboration with your team and with visible support from your supervisor or manager.

Key questions:

- Why the panel?
- What type of panel? And why?
- What will be the visible outcomes from the panel?
- Will it be a one-time only, or repeat?
- Who will pay?

2 Assign a Facilitator

You will want a facilitator or point person who has the skills and time to deliver results.

This person must understand the objectives of the panel in order to lead the group.

See Also: Charter for a Customer Panel

Important: It is best to form a core team of three to five people who can facilitate the planning and execution of your panel. Team members are best. In larger corporations your travel department, event planning and other support functions can help.

3 Select the Core Team

Key questions:

- What skills will they need?
- Do they have the time to participate?
- Do they have the support of their managers or supervisors?

Be sure to prep your core team members for the tasks.

Important: Attention to logistics is absolutely critical. "No loose ends" must be your strategy. Develop a process map or a Work Breakdown Structure (WBS) to ensure accuracy. Over-communicate!

4 Develop a Discussion Guide

Key questions:

- What do we need to find out?
- How are we going to find it out?
- What is the natural flow to find it out?
- What will we do with what we find out?

See Also: Discussion Guide

5 Profile the Customers

Key questions:

- Are they representative of the customer segment chosen?
- Do we have a true cross-section of the targeted customers?

Continued

Customer Panels

CONT'D

6

RECRUIT THE CUSTOMERS

Key questions:

- Will the customers have the time and desire to participate?
- Will the customers be willing to participate more than once?
- Will there be an incentive to help motivate participation?
- Is the selected venue conducive to participation (i.e., conveniently located)?

7

COORDINATE ALL LOGISTICS

In addition to on-site logistics such as venue, refreshments, clearance, etc., consider logistics such as travel arrangements for you, your team and your customers. Maybe lodging too.

8

HOLD THE CUSTOMER PANEL MEETING

9

EXTENDED MEETING AFTER THE PANEL

Key questions:

- What did we learn?
- What was the true message from the customer?
- What specific action will we take as a result of the panel?
- What will we incorporate in our next product?

Important

Conduct this meeting immediately following the panel. One benefit, you capture information while it is still fresh. Also, you can meet with your team without the typical office interruptions. We suggest a two- to three-hour meeting for every panel.

10

DEVELOP AN ACTION PLAN

Key questions:

- What did we learn during the panel?
- What specific action must we take as a result of the panel?
- Who will be responsible for the actions?

Important

Remember to thank your customers. It is the civil thing to do, and it opens the door for future interaction.

Example: Charter for a Customer Panel

1. What is the purpose of the Customer Panel? Why are we holding this meeting? For what intent?

2. What are the objectives? What is the measurable accomplishment from the meeting?

3. What are the tangible results or outcomes that we must have from this panel?
   For example:
   - A description of the process that customers apply to our product.
   - A prioritized set of needs from the customer.
   - Understand how change is impacting the use of our product.

4. What are the methods that we will use to achieve the objectives? What is the overall approach?
   For example:
   - Create a very detailed agenda.
   - Ensure full participation by all team members.
   - Divide into groups to identify barriers and obstacles.
   - Create a Core Team to steer the program to completion.

5. What are the roles for the panel?
   For example:
   - Who is the facilitator?
   - Who will coordinate logistics?
   - Who will take notes?
   - Who will create the handouts?
   - Who will make presentations?

6. What will happen the day after the panel?
   For example:
   - All notes will be transcribed.
   - A meeting of the Core Team will be scheduled.
   - All action items will be sent to the attendees.
   - Who needs to know the results of the meeting?

7. What will happen a week or two after the panel?
   For example:
   - The Core Team will meet.
   - Various sub-teams will have meetings.
   - A presentation will be made to the executives.

8. How will we evaluate the success of the panel?
   For example:
   - Debrief the attendees.
   - Create an evaluation form.
   - Assess the number of deliveries that we complete.

Important

Use this form as either a checklist or a template where you write in the information.

This form should not be filled out by an individual, but rather it should be a group exercise used to engage your team.

Customer Panels

9. Who from our organization must be part of this panel?

- ❖ Engineering.
- ❖ Marketing.
- ❖ Manufacturing.
- ❖ Technology (R&D).
- ❖ Quality.
- ❖ International.
- ❖ Suppliers.
- ❖ Channel management.
- ❖ Sales Management.
- ❖ Documentation.
- ❖ Advertising and PR.
- ❖ Company Executives.
- ❖ Consultants.
- ❖ Motivational speaker.
- ❖ Training.
- ❖ Customer support.
- ❖ Service and repair.
- ❖ Calibration and metrology .

10. Logistics?

For example:

- ❖ How long will the meeting be?
- ❖ Where will the meeting be held?
- ❖ What is the starting time? Are there multiple starting times?
- ❖ What is the ending time?
- ❖ What about meals, refreshments and other amenities?
- ❖ What will be the layout of the room?
- ❖ Will we need more than one room for the meeting?
- ❖ Lodging.
- ❖ Travel.
- ❖ Audio-visuals such as projectors.
- ❖ Flip charts.
- ❖ Internet connection.
- ❖ Evening events.
- ❖ Other events.
- ❖ Office supplies.
- ❖ Printers, faxes.
- ❖ Who will pay?
- ❖ Does our development schedule allow time for the meeting and its impact on the new product?

Appendixes

Recommended Books

Introduction

These lists reference books that the senior consultants at Rapidinnovation, LLC have read and found applicable to today's business challenges. Rapidinnovation, LLC is a consulting firm whose primary focus is serving product development organizations around the world, which we've been doing since 1997, and with many satisfied clients.

We welcome your recommendations and reviews of these and other books and resources. Please e-mail the authors with your comments and suggestions.

Teams and Teamwork for Product Development

Book Title & Author	Subject	Comments
The Wisdom of Teams By Jon R. Katzenback Published by Harvard Business School Press	• High-performance teams.	• Excellent book for creating development teams. • Clear, no-nonsense approach to cross-functional teams such as concurrent engineering, product development, etc.
Peopleware, Productive Projects and Teams, Second Ed. By Tom DeMarco & Timothy Lister Published by Dorset House Publishing	• Software development. • People management.	• Good book to help understand the human side of software development.
Principle-Centered Leadership By Stephen R. Covey Published by Fireside	• Timeless principles of leadership. • Clear, no-nonsense understanding of leadership in a business environment.	• The difference between a manager and a leader, and its implications in an organization.
Punished by Rewards By Alfie Kohn Published by Houghton Mifflin Company	• Reward, recognition, celebrations and motivation in organizations.	• The truth about failed recognition programs and the unintended results of poor motivation and recognition programs.
Mastering Virtual Teams, Third Ed. By Debora L. Duarte and Nancy Tennant Published by Jossey-Bass	• The ways and means for enabling virtual teams to succeed.	• Duarte does a good job of presenting the challenges and solutions for distributed teams. A level-headed presentation of the issue.

Recommended Books

General Management and Strategy

Book Title & Author	Subject	Comments
The Innovator's Dilemma By Clayton M. Christensen Published by HarperBusiness	• Why companies fail to stay at the top of their industries.	• A book that confronts a reality faced by most successful technology companies: the need to change, and change fast.
The Discipline of Market Leaders By Michael Treacy Published by Perseus Books	• Choose your customers, narrow your focus, dominate your market.	• Deliberately choosing the customers you wish to serve allows you to develop the competencies needed to succeed.
Built to Last By James C. Collins Published by HarperBusiness	• Visionary companies. • Values, culture and successful businesses.	• Focusses on the human side of corporate culture for the promotion of long-term success.
The Balanced Score Card By Robert S. Kaplan and David P. Norton Published by Harvard Business Press	• What to measure and why. • Financial measures alone are not enough to ensure sustainable customer satisfaction and profits.	• Rational approach to measures. • Systemic look at the enterprise. • Focus on the human side of measures.
Companies with a Conscience, Third Ed. By Mary Scott and Howard Rothman Published by Citadel Press	• Case studies of 12 companies that are profitable and at the same time people-, environment- and community-friendly	• Good examples of "having your cake and it eating it too" — the ability to make a profit and have a clear corporate conscience.

Product Development

Book Title & Author	Subject	Comments
Rapid Development: Taming Wild Software Schedules By Steve McConnell Published by Microsoft Press	• Perspectives and methods for rapid software development. • Focus on the process and framework for development.	• The methodical approach to rapid software development. • Good food for thought.
Revolutionizing Product Development By Steven C. Wheelwright Published by The Free Press	• One of the classics on product development.	• A must-have for those who are serious about product development.
The Invisible Computer By Donald A. Norman Published by The MIT Press	• Why good products can fail.	• Insightful as to why technology companies behave like technology companies.
Dynamics of Software Development By Jim McCarthy Published by Microsoft Press	• Rapid software development is not the result of countless hours of coding.	• A book more about teams and teamwork than software development.

Recommended Books

Book Title & Author	Subject	Comments
Portfolio Management for New Products, Second Ed. By Robert G. Cooper, Scott J. Edgett and Elko J. Kleinschmidt Published by Perseus Books	• Managing of technology for profits. • Disciplined product development. • Separate the losers from the winners.	• A solid, direct book on managing a portfolio of products while optimizing resources. • Good methods for balancing your product offering.

Productivity Improvement

Book Title & Author	Subject	Comments
The Fifth Discipline By Peter M. Senge Published by Doubleday	• System thinking. • The quintessential book of systemic approaches to solve business problems.	• A must-read to understand the total solution to customers' problems • Also very important to re-engineering
Improving Performance: How to Manage the White Space on the Organization Chart By Geary A. Rummler Published by Jossey-Bass	• Business processes. • Teamwork in organizations. • Process mapping for non-manufacturing organizations.	• Clear understanding of processes in a business environment. • Good how-to information and a level-headed approach to organizational development.
Kaizen: The Key to Japan's Competitive Success By Masaaki Imai Published by McGraw Hill	• Continuous improvement.	• The truth about continuous improvement as a culture and not a fad.

Project and Program Management

Book Title & Author	Subject	Comments
Flexible Project Management for Product Development By Jose Campos Published by Rapidinnovation, LLC	• Clear understanding of the conventional project management rules and tools do not necessarily work in a product development environment. But here, the author has experience in both project management and product development.	• Easy to understand and very detailed. • Plenty of examples, templates and forms ready to use. • Easy to share with other members of the development team. • Clearly written with product development in mind, very practical.

Recommended Books

Book Title & Author	Subject	Comments
Project Management Toolbox By Dragan Milosevic Published by John Wiley and Sons	• A comprehensive set of project management tools full explanations and plenty of examples	• A must-have reference
Program Management for Improved Business Results By Dragan Milosevic, Russ Martinelli and James Waddell Published by John Wiley and Sons	• A business approach to the role of program managers. Authored by Dragan Milosevic, acclaimed authority in project and program management.	• Probably the best book on program management.
Reinventing Project Management By: Aaron J. Shenhar Published by Harvard Business Press	• Proof positive that every project is different and therefore, you cannot manage them exactly the same.	
The Fast Forward MBA in Project Management, 2008 Ed. By Eric Verzuh Published by John Wiley and Sons	• Looks at the challenges of today's project environment and examines the skills a project leader needs to make a project successful. • It includes organizational changes, tools and techniques, control methods and real-life applications.	• Good medium-depth overview of all aspects of project management. • Easy to read, with examples and case studies, but meaty enough that both junior and experienced project leaders will find nuggets. • Down-to-earth
The New Project Management, Second Ed. By J. Davidson Frame Published by Jossey-Bass	• Frame makes a case that conventional wisdom is only marginally relevant to project leaders working in today's turbulent project environments. • He offers strategies and techniques for changing from implementor to initiator, and for dealing with the soft and complex issues that affect most technology projects.	• Good ideas, especially on soft skills and orientation toward the customer. • Frame sometimes undervalues the traditional techniques and gets too enthusiastic about how empowered a project leader can be. Pair this book with a solid grounding in the hard fundamentals of project management, such as from Verzuh's book.
Visualizing Project Management, Third Ed. By Forssberg, Mooz and Cotterman Published by John Wiley and Sons	• A recently updated project management textbook with lots of examples. It strives to provide an integrated model of project management, and focuses on vocabulary, teamwork, the project cycle and management techniques.	• Covers a wide variety of hard and soft project management topics, with short case studies of how they have been used in real situations. • Focuses mostly on applying them within a sequential framework, so don't look for lots of insight on iterative approaches here. The authors' experience in aerospace and government projects shows through in the vocabulary and examples.

Recommended Books

Book Title & Author	Subject	Comments
The Wiley Guide to Managing Projects By Peter W. G. Morris and Jeffrey K. Pinto Published by John Wiley and Sons	• Published in 2005, this book is up-dated every 10 years. This is a comprehensive book on project management. Looks well beyond the basics on project management (cost, schedule and performance).	• A must-have reference for those who are serious about project management. • Critically important for those involved in the strategies of project management.

The Voice of the Customer (VOC)

Book Title & Author	Subject	Comments
The Voice of the Customer for Product Development By Jose Campos, Jean-Claude Balland Ph.D. Published by Rapidinnovation, LLC	• A detailed guidebook on how to capture and use the voice of the customers (VOC) for product development. The author describes every step and detail in an easy to understand fashion.	• Easy to understand and very detailed. • Plenty of examples, templates and forms ready to use. • Easy to share with other members of the development team.
How Customers Think By Gerald Zaltman Published by Harvard Business School	• Essential insight into the mind of the customers and the market.	• Customer needs are much more subjective than we thought. • An eye opener concerning value.
Advanced QFD By Larry Shillito Published by Wiley & Sons, Inc.	• The use of QFD (Quality Function Deployment) in a real-world environment, linking technology to the market.	• A sound, levelheaded book on QFD, A must-have if you are serious about QFD.
Software Requirements, Second Ed. and More About Software Requirements By Karl Wiegers, Published by Microsoft Press	• A practical guide for gathering and managing requirements throughout the product development cycle.	• As relevant to product requirements as it is to software requirements, this practical guide will go a long way towards making your R&D and Marketing people see the world through the same lens.
Customer-centric Product Definition By Sheila Mello Published by AMACOM	• Good coverage of the Voice of the Customer (VOC) and its use in defining new products.	• Very good book; the author clearly portrays the process of getting the customers at the center of new product development.
Writing Effective Use Cases By Alistair Cockburn Published by Addison Wesley	• A practical guide to the powerful approach of expressing the behavioral requirements of complex systems.	• Practical, with numerous examples of good and bad use cases.
Customer Visits, Second Ed. By Edward F. McQuarrie Published by Sage Publications	• A clear description of what it takes to perform a good customer visit.	• The author is well-recognized as an authority in this field. Levelheaded approach.

Recommended Books

Book Title & Author	Subject	Comments
Acquiring, Processing, and Deploying the Voice of the Customer By M. Larry Shillito Published by St. Lucie Press	• Description of ways to obtain, process and use the voice of the customer.	• Plain English, down-to-earth, but clearly from someone who knows.
Voices Into Choices By Gary Burchill Published by Joiner	• A handbook with step-by-step descriptions to obtain and use the voice of the customer	• Comprehensive. • Truly a fine work — a must-have handbook.
Priceless: Turning Ordinary Products Into Extraordinary Experience By D. LaSalle & T. A. Britton Published by Harvard Business Press, 2003	• The title says it all. People don't buy products, they buy value and experience. This book will help you see the world through the eyes and heart of customers.	• Very valuable in bringing the new concept of experience to a tactical level.
Quality Function Deployment: How to Make QFD Work for You By Lou Cohen Published by Addison-Wesley, 1995	• A tour of QFD fundamentals, as well as advanced concepts	• Clear exposé to a sophisticated subject. • Most useful to engineers than marketers.
The Observational Research Handbook: Understanding How Consumers Live with Your Product By Bill Abrams Published by McGraw-Hill Professional, 2000	• Practical coverage of the art of observing customers in context to understanding latent needs.	• Useful advice on how and when to observe customers to get clues to their needs.
Learning from Strangers: The Art and Method of Qualitative Interview Studies By Robert S. Weiss Published by Simon & Schuster, 1995	• The subtitle of the book says it all: "The art and method of qualitative interview studies." These are important skills to master.	• A how-to book on interviewing customers by an expert on the topic.

Recommended Books

Risk Management and FMEA (FMCA)

Book Title & Author	Subject	Comments
Risk Management and FMEA for Product Development By: Jose Campos Published by Rapidinnovation. LLC	• A detailed guidebook on risk management for product development. The author describes every step and detail in an easy to understand fashion.	• Easy to understand and very detailed. • Plenty of examples, templates and forms ready to use.
Waltzing with Bears: Managing Risk on Software Projects By Tom DeMarco and Timothy Lister Published by Dorset House Publishing		
Proactive Risk Management: Controlling Uncertainty in Product Development By Preston G. Smith and Guy M. Merritt Published by Productivity Press	• A "plain English" book on risk management.	• A good reference for anyone responsible for managing risk.
Managing the Unknown By Christoph H. Lock, Arnound DeMeyer and Michael T. Pich Published by John Wiley & Son	• A book about managing the risks you don't know: the ways and means to deal with total uncertainty. • Intriguing new concepts about risk, particularly in areas with high uncertainty such as new technology.	• A good book to have if you are in product development or technology development. • Challenges conventional views of risk management.
Project and Program Risk Management: A Guide to Managing Project Risks and Opportunities By R. Max Widerman Published by Project Management Institute, 1992	• A risk management book for beginners and intermediate managers of risk.	
Failure Mode Effect Analysis: FMEA from Theory to Execution, Second Ed. By D. H. Stamatis Published by American Society for Quality, 2003		• A comprehensive book on FMEA, strong in context and advanced methods. • A good overall reference book for those serious about FMEA.
Project Management Tool Box By Dragan Milosevic Published by John Wiley and Sons, 2003	• A comprehensive set of project management tools, full explanations and plenty of examples.	• A must-have reference.

Glossary

APPENDIX B

Definitions for Product Development

Practical Definitions in Plain English to Enable Effective Action

The intent of this glossary is to explain the terms you will encounter in the real world of product development. Furthermore, we explain these terms in a practical, common-sense fashion. When multiple relevant definitions exist, we will provide them. With this resource, you will be able to better connect the concepts and tools of product development to your projects.

Term	Definition & Notes
Agile	• A software development model or framework based on frequent iterations with the customer and development of the software in small, incremental amounts, collaboration and flexibility.
Alpha Unit	• A version of a new product that is ready for internal testing. The product meets criteria that enable testing that will yield valuable data and input to proceed with the final version. • After the Alpha Unit comes the Beta Unit, a more mature version of the product. • In hardware an alpha unit may be used to test the manufacturing process, functionality or testing strategy. • In software it can be used to test human interface (GUI), usability or functionality. • Alpha units are not saleable and are intended for internal use only.
Assumption(s)	• A short set of relevant factors that can have a positive or negative impact on your project or program and define the playing field. – The playing field refers to the arena or area of focus for your endeavor. – A short set means three to seven relevant factors. – Relevant means of importance to you and your team — that which must be taken into consideration when making decisions.
Attribute	• Used to define a set of general characteristics of a platform, product line or product, an aggregate of several features. Examples include "ease of use," "ruggedness" and "re-configurable". • Related to, but not the same as "Product feature." See also "Feature."
Awareness	• A measure of the percent of target customers who are aware of the existence of something. It could be a product, a brand or even a concept. • Awareness does not imply preference. • Awareness can be measured for brand, product or any other attribute you select. • Awareness should be measured before and after an event; for example, an advertising campaign. The difference provides a measure of effectiveness. • There are two kinds of awareness: Unaided awareness is when the customer volunteers the information (manifests the awareness) without any help. Aided awareness is when a hint or suggestion is provided, such as a photograph, and then the customer expresses awareness.

Glossary

Term	Definition & Notes
Benchmarking	• The process of studying how successful companies, industries and even competitors perform specific functions and then selecting the best as standards of performance for your own improvements. • You can benchmark with your competitors or across industries, in your own country or across the world. • Benchmarking is a good way to develop your level of excellence, and to learn "good practices" for your continuous improvement.
Best Practices	• A method, technique, tool or other practice that has shown to be the best within a certain category. Best practices must be researched, defensible and published by an authorized organization. • Methods, approaches or processes deemed to be "best in class"and, consequently, worthy of imitation. • See also "Good Practices."
Beta Test (a.k.a. Beta Unit or Beta Version)	• An external test of pre-production products. A way to test new products in actual application to discover bugs and last minute fixes before full introduction. • Some companies (software in particular) use Beta Tests as part of their introduction strategy, to generate awareness and word of mouth.
Bill of Material (BOM)	• A hierarchical list of systems, sub systems, sub assemblies and components that make up a product or a higher level sub-system. • It can be a set of parts list, drawings, and schematics. In manufacturing it can also mean the way a new product is assembled including all parts, systems and sub systems.
Brainstorming	• The process of tapping into a group's creativity to generate ideas, solutions or innovation. • Generally, it takes the form of a meeting or gathering where the team or group is encouraged to come up with as many creative ideas as possible. • There are many methods to facilitate or enable brainstorming. The methods are intended to enable the ideation, processing of the information, and finally prioritization of the ideas. Many brainstorming sessions fail due to a lack of a framework or process.
Brand	• The combination of all tangibles, intangibles and actions of a company that create a particular promise of value in the customers' mind. • A brand is a holistic concept that evokes the mental images and expectations the customers have about a particular company.
Breakeven	• A financial metric that identifies the point where the revenue (sales) of a new product match the investment made in developing it. • At this point, the new product has reached enough sales to start generating "positive cash flow."

Glossary

Term	Definition & Notes
Business Case	• The results of the business analysis to demonstrate the viability of a new product or service prior to approval for development. • Viability includes profitability, market share or other critical and strategic objectives of your company. • "Making the business case" in product development implies your ability to convince the company executives that investing in your new product will result in a measurable return. Philosophically, you are going to the company executives to ask for an investment, which will results is a quantifiable return
Business Model	• The ways and means selected by the company to generate revenue and make a profit. • The Business Model is said to be working when the company has deployed and executed the strategies to generate revenue, profits and serve its customers.
Business to Business (B-to-B)	• Sales to other businesses and not to individual consumers. For example, an OEM sells components to a manufacturing company. • Sales to distributors, representatives or wholesalers are also considered B-to-B.
Champion	• A person who takes strong personal interest in seeing that a particular process, product or other is carried to fruition. Champions are always ready to deal with adversity and committed to success. • Champions are said to be "maniacal" in their commitment, willing to put exceptional effort and energy into their cause. • No innovation can be brought to fruition without champions.
Change Management	• An organizational process aimed at helping employees embrace changes in their current business environment and then change their behavior in support of the new state. • A structured approach to shifting or transitioning individuals, teams and organizations from a current state of behavior to a desired future state.
Charter, Team Charter	• The document that officially starts an initiative. This document outlines the team members, the common purpose that brings them together and the performance goals to ensure progress and outcomes. It also defines the roles, responsibilities and authority. It defines the desired approach to ensure completion. • The team members should be strongly involved in — even responsible for — the development of the charter. • The Charter must be fully supported by Management. • A charter is the unifying principle of the team. Without it, each team member has a different idea of what needs to get done and how.
Co-location	• Product development where the entire development team is in one location; for example, the same building. • The processes and methods for co-location development are different from that used for multi-site or distributed product development.
Commercialization	• The process of taking a new product to market. The transformation of an idea into a product that customers are willing to purchase. • Commercialization can be used for a new product, a new technology or anything that a particular set of customers value and are willing to pay to obtain it.

Glossary

Term	Definition & Notes
Compelling Reason To Act (CRTA)	• The needs the customer must address now so that a budget and the will to act can be generated. • CRTA is proposed to designate the highest priority needs — the ones that can be funded and will be acted upon by your customer. • A customer must have a CRTA in order to purchase your product. It is your responsibility to identify CRTAs so you can develop products that address customer requirements.
Competency	• The aggregate set of skills, knowledge and abilities that an individual needs to fulfill a specific set of measurable responsibilities. • Competency is all about one individual — not a team or an organization. The equivalent statement at the organizational or team level is known as "maturity" or "organizational maturity"
Complete Product	• The aggregate of the tangible, intangible and emotional components that constitute a product in the eyes of the customer. • See also "Offering" and "Whole Product."
Completion	• The verifiable fulfillment of all objectives. • The conclusion of all efforts; for example, the completion of a new product development is the introduction or rollout to the market place.
Compound Annual Growth Rate (CARG)	• The calculation of rate at which an investment would grow if it grew at a steady rate. It's a way to smooth the return on an investment for the purposes of making informed decisions. • In product development, CARG is generally used to describe the growth rate of a market or the growth rate of new products.
Computer Aided Design (CAD)	• The use of computers and software to design a new product or parts of a new product. • CAD is a very advanced technology enabling simulation, testing, and all aspects of design.
Computer Aided Manufacturing (CAM)	• The use of computers and software to design and run a manufacturing process in an integrated fashion; that is, as a system. • In some areas CAM is also used to describe any aspect of the manufacturing process that is run with the aid of computers or fully automatic, not necessarily the entire system of manufacturing. • Widely used in industry, CAM has become essential to running manufacturing operations around the world. • CIM (Computer Integrated Manufacturing) may be used in conjunction with CAM.
Computer Integrated Manufacturing (CIM)	• See "Computer Aided Manufacturing (CAM)"
Concurrency	• The ability to develop several products at the same time. • The goal of a product company is to achieve speed, concurrency and repeatability. • It can also be applied to the development of a single product where parts, assemblies and subassemblies are developed at the same time and not sequentially.

Glossary

Term	Definition & Notes
Concurrent Engineering	• A model of product development where all functions such as engineering, marketing, manufacturing, etc., work at the same time to design and take the product to market. • A set of methodologies that enable rapid product development by allowing as many activities as possible to take place simultaneously. • Contrasted to the sequential model, where one activity had to wait for another before it could be started.
Concurrent Engineering	• An emerging engineering methodology based on the premises: – All aspects of the life-cycle of a product should be considered as part of the initial design process; that is, from basic functionality until end of life. – All activities that can happen at the same time must be undertaken. • The methodology, if properly applied, yields better products in a shorter amount of time.
Conjoint Analysis	• A quantitative market research technique used to characterize how people value the features (or benefits) of a product. • One of the few methods available to determine the possible price of a new product with objectivity. • Conjoint analysis is powerful because it puts the customer in a real life situation of having to make trade-offs between benefits (what he wants to get) and costs (what he has to give to get it).
Constraint	• Restriction or boundary imposed on a project or design. • Constraints may help in setting priorities, avoiding risks or the physical limitation of the technology.
Contextual Inquiry	• A way to gain knowledge about a customer by observing people in their environment doing specific activities. • The use of anthropology, social science and other techniques to study customers performing specific tasks for the purpose of understanding their needs. • The observation must be qualitative.
Core Competency	• The few things that a company does that: – Its customers find exceptionally valuable. – It does better than its competitors. – Its competitors find difficult to imitate. – It would never outsource to anyone. – It can transfer to other industries as is. – Is part of its growth plan. • The definition of Core Competency varies. It's driven by the specific situation of a company at the time it defines its Core Competencies. Nonetheless, it is critical for a company to understand and develop its core competencies. • Core competencies are always a web of things that a company does particularly well without sometimes even be conscious of it.

Glossary

Term	Definition & Notes
Core Team: Concurrent Engineering	• A small group of representatives from the critical functions (e.g., Engineering, Marketing, Program Management, Finance, etc.) responsible for the development of a specific product and accountable to deliver measurable results. • The Core Team is responsible and accountable for progress and eventual delivery of the new product to the market. It acts as the Program Manger for the development.
Core Team: Generic	• A small group of three to five individuals that is chartered and empowered to guide the implementation and completion of an initiative. • Very useful in fast-changing environments where distractions abound because a small group can focus attention and energy to ensure clean execution.
Corrective Action	• The specific steps to eliminate the cause of a detected nonconformity or other undesirable situation. That it, the steps that are taken to permanently remove the causes of an existing nonconformity through root cause analysis and deploying permanent corrections. • In contrast, "Remedial Action"refers to the temporary steps taken to eliminate one specific non-conformance without regards to a permanent solution.
Criteria	• A standard used to evaluate alternatives in a decision. Criteria are always finite and easy to understand by all. • In the product development process, the criteria are generally defined by the Gatekeeper at major gates.
Critical Path (CP)	• The path with the longest duration • CP comes from the analysis of a network diagram or a PERT (Project Evaluation and Review Technique) diagram in project management.
Critical Success Factor (CSF)	• Something that is critical to your program and must happen. • The few key areas where things must go right for the business to flourish. If results in these areas are not adequate, the organization's efforts for the period will be less than desired. • Those few things that must happen in order to achieve the desired results.
Cross Section (Market)	• A typical selection or sample. For example, a cross section of a particular market segment. • This term is a metaphor from engineering, used to reflect the framework of a particular structure such as a market segment.
Cross-Functional Team (a.k.a Multi-Functional Team)	• A team made up of representatives from various functions. For example, marketing, engineering and manufacturing. • Most product development is achieved though cross-functional teams.

Glossary

Term	Definition & Notes
Culture	• A mosaic of interrelated elements and their collective day-to-day interaction that shape the behaviors of people in the group and what's right and wrong. • For example: – Management behavior. – Customs, myths and norms. – Ceremonies, rituals and events. – Rewards and consequences. – Physical environment. – Rules and policies. • A company culture develops when corporate values are consistently followed for a number of years. • Said another way: "Culture Consists Of A Shared, Commonly Held Body Of General Beliefs And Values That Define The 'Shoulds' And The 'Oughts' Of Life." — Kluckhohn
Customer	• The person or entity that pays for the value of a product or service. • There is tendency in some organizations to refer to teams and functional groups as "internal customers." This tends to diffuse the importance of the one and only true customer, the one that pays. • Customers can be "users" or not of your product. Make the distinction between the person who pays and the person who uses.
Customer Experience	• A succession of events in the life of the customer dealing with an offering. This interaction will generate good or bad feelings. • Designing experiences is the new frontier in user-centered design because creating pleasant feelings is the best way to generate satisfaction, loyalty and word of mouth. • Thinking in terms of customer experience forces the marketer to consider the entire sequence of events that the target customer undergoes when dealing with a company and its offerings.
Customer Intimacy	• The business definition as introduced by Michael Treacy in Discipline of Market Leaders is (paraphrase): The discipline of delivering what each specific customer wants or needs rather than what the market as a whole wants or needs. • Customer-intimate companies develop deep relationships with their customers so as to anticipate their needs and solve their problems better and faster.
Customer Requirements	• The crystal-clear articulation of the desired outcomes a customer is striving to achieve.
Customer Satisfaction	• The positive or negative evaluation or feeling of customers that results from comparing the actual experience with a product or service with the expectations it had prior to purchase and use. • Customer satisfaction is a different concept than "customer value." A customer may be satisfied or not with the value it gets.
Customer Value	• The perception of what customers will get (benefits) for what they have to give (costs). • Alternatively, it is the customers' perception of the net benefits they will derive from your offering at a particular price, situation and use.

Glossary

Term	Definition & Notes
Customer Value Drivers (CVDs)	• The criteria that are perceived by the customer to be the most important to the buying process. The yard stick that the customer uses to make the buy decision. • Customer value drivers can be tangible (objective) or intangible (subjective).
Customer-Driven	• The urge to grow a business culture committed to the relentless creation of superior value for the customer. • It is a set of values that put the customers' interest first in order to develop a long-term profitable enterprise — while not excluding those of other stakeholders (paraphrased from: Narver and Slater). • In a for-profit business it is part of the overall strategy to generate revenue and profits. A company can make a profit and not be customer-driven.
Defect	• Lack of conformance to a standard. • Since the introduction of Quality methods, a defect is anything that your customer says it is. This is an effort to focus the organization to evaluate products and services as the customer would.
Defects: Engineering	• Engineering "defects" include design errors, software bugs, incorrect documentation and any substandard design or development steps that would not occur if the process were under control.
Deliverables	• The visible or "touchable" results from a task or a project. The outcomes from a task or a project. • Deliverable has come to mean more than "able to be delivered", now more at "the results of an endeavor or a project." • For example, the deliverable from the marketing organizations is a Marketing Requirements Document (MRD).
Delphi Process	• A technique that uses iterative rounds of consensus development across a group of experts to arrive at the most probable outcome for some future state. • "Experts" are people who have knowledge in a particular area; for example, knowledge about a specific market segment.
Derivative Product	• A new product based on changes to an existing product without affecting the basic product architecture or platform. • A derivative product can be an improvement to an existing product or a different version to satisfy a specific set of customers. • In software development, a rev includes new value to the customer. A rev to fix bugs is not a derivative product.
Design Change Order (DCO)	• A process to ensure the smooth and controlled change of a design and minimizing disruption. • Documented changes to the design of a new process, which is part of a formal process for changes. • The purpose of the DCOs is to minimize disruptions to the design process and ensure clarity of communication. • Also used in this context: – ECN: Engineering Change Notice – ECC: Engineering Change Control – ECO: Engineering Change Order

Glossary

Term	Definition & Notes
Design Engineer	• A person with proven domain expertise and experience who is responsible for creating innovative solutions that customers value. • The output of a design engineer is a package of technology in the form of documentation (drawings) that is the tangible manifestation of an innovative solution that customers value.
Differentiated Product	• Characterizes a product that provides more advantages than its competitors for the most important Customer Value Drivers of a given segment. • A well-differentiated product is one of the most important success factor for a new product.
DOA (Dead on Arrival)	• Products that arrive at a customer site unusable. • A product that does not work when the customer unpacks it.
Early Adopters	• Customers who buy into a new product or technology very early in its life cycle. • Early Adopters are willing to take the risk of an unproven new product or technology to gain a business or personal advantage.
Economic Value Analyses (EVA)	• A methodology that enables you to understand the real progress of a development process by measuring actual progress towards completion. A way to clearly understand progress towards the objectives of a new product. • EVA is found in project management texts — very popular in the construction industry — and now being adopted in product development due to its objectivity. • While budge and schedule are indicators of "movement", they do not necessarily show progress towards the achievement of the objectives of the new product.
Elevator Pitch (a.k.a. 30-Sec. Commercial)	• A short statement summarizing in a compelling way a value proposition. • If a pitch can't interest a prospect in the time it takes to move a few floors in an elevator, you don't have a strong, well-differentiated offering. Rework the pitch and try again.
Engineering Change Control (ECC), **Engineering Change Notice (ECN),** **Engineering Change Order (ECO)**	• A formal process that documents a needed design change regardless of the root cause of such change. The process must follow sound engineering principles for control, tracking and disposition. • See "Design Change Order (DOC)." • EISs may include modeling, simulation, analysis and other computational requirements.
Engineering Information System (EIS)	• Applications and solutions for information-intensive engineering and scientific endeavors. The ability to model, process and manage very large amounts of information for the purpose of enabling engineers and scientists to accomplish their goals.
Experience	• A succession of events in the life of the customer dealing with an offering. This interaction will generate good or bad feelings. • Designing experiences is the new frontier in user-centered design because creating pleasant feelings is the best way to generate satisfaction, loyalty and word of mouth. • Thinking in term of customer experience forces the marketer to consider the entire sequence of events that the target customer undergoes when dealing with a company and its offerings.

Glossary

Term	Definition & Notes
Failure	• A defect, nonperformance or lack of adherence to a specific requirement. • Something that causes a product to no longer fulfill its intended function.
Failure Analysis	• A process to methodically characterize and determine the root cause of a failure. The reasoning from the symptoms of a failure to the identification of the root cause of the failure. • Used in reliability engineering, as well as FMEA (Failure Mode and Effects Analysis).
Feature Creep	• The tendency to add new features and attributes to a new product during the design process. • Feature Creep is not a desirable activity as it delays the product, adds to its complexity and may even interfere with the business objectives of the new product. It also increases the stress and aggravation of the entire development team.
Feature (a.k.a Attribute)	• A quality or capability of an offering. • When they are familiar with the product category, customers often express their needs in term of features or attributes (e.g., "I need a 4-door sedan that gets more than 25 mpg") making it more difficult to understand their primary needs.
Firefighting	• The act of addressing too many interruptions, trivial issues and other distraction at the expense of the schedule. • Taking care of surprises and other unexpected distractions. • Firefighting is generally caused by a severe lack of planning and the lack of a risk management plan.
FMEA (Failure Mode and Effects Analysis)	• A structured process to identify potential failures of new products, components or subsystems. A form of risk management. • Can also be used to analyze and determine the root cause of actual failures.
Focus Groups	• A market research tool where a qualified facilitator (moderator) leads a discussion with a select group of target customers for the purpose of understanding their needs and preferences. • Focus groups are generally eight to twelve customers in a single room; generally the room is designed for focus groups. The facilitator (moderator) leads the conversation, and the response is observed and also video taped for later analysis. • Focus groups are a qualitative form of market research.
Functional Test	• Testing a circuit board to identify functional level faults or defects.
Funnel	• See "Roadmap."

Glossary

Term	Definition & Notes
Fuzzy Front End	• The beginning of a program or project when much is not known and a high level of confusion exists. It's also the best opportunity to influence the long term success of the program or project. All programs and projects have this phase. • Initially, this term was defined as the period of time when the development of a new technology was not clear (fuzzy). The meaning has been extended to include the start of a product development project when much confusion exists — more questions than answers. • We suggest that every development project has a fuzzy-front-end; consequently, it is best to acknowledge it and set about exiting this phase by obtaining the missing information
Gantt Chart	• A horizontal bar chart used in project scheduling that shows the start and end dates of tasks within a project. • A Gantt Chart is a very common way to depict project schedules due to its simplicity and visual impact.
Gate	• The decision point, generally a meeting, at which a management decision is made to allow the product development project to proceed to the next gate. • A gate has only three outcomes: "YES", "NO" or "Delay." • Gates are generally related to the timeline of the project; consequently a gate has a chronological impact. • The metaphor of a gate is part of the Gate Phase approach to project management, which has been part of product development process for many years. This approach is now more likely replace by "Agile" and other forms of iterative product development.
Gate Phase	• A product development process or methodology based on specific stages, where at the end of one stage and prior to the start of the subsequent stage there is a review to ensure that a set of criteria has been met.
Gatekeeper	• The person who or organization that holds approval. • In product development, the person who or organization that decides if the product development team can proceed with a new stage if the criteria from the previous stage has been met.
Go/No Go Gate	• A gate in the development process that allows only two outcomes: proceed or stop. • Critical in any development process, as it determines if the development meets the criteria to proceed to the next gate in the process. • A "no go" decision does not necessarily mean end of the development project.
Goal	• A purpose toward which a project is aimed. • A general direction of achievement that is not measurable. Rather, it is statement of intent; for example, "increase sales" or "accelerate product development." • We distinguish between "goal" (immeasurable achievement) and "objective" (measurable achievement). Some choose to reverse these terms.
Good Practices	• Obtained by establishing benchmarks to identify what meets a criteria of excellence. • See also "Best Practices."

Glossary

Term	Definition & Notes
High-Performance Team	• A small number of people with complementary skills who are committed to a common purpose, performance goals and approach for which they hold themselves mutually accountable — and are deeply committed to one another's personal growth and success. — from The Wisdom of Teams, by Katzenback.
Hurdle Rate	• The minimum return on investment or internal rate of return a new product must meet or exceed as it is considered for approval and proceeds through development. • A Hurdle rate can be applied to any investment.
Ideation	• Similar to brainstorming but focused on generating specific ideas about improving the performance of a product or service. • The process of idea generation generally occurs at the early stages of development.
Inbound Marketing	• The collection of all activities that provide information about the market, including market research, customer research, feedback from other organizations like sales, training and support, industry reports, competitive analysis and technology assessments. • The function responsible for obtaining and prioritizing customer requirements, then collaborating with engineering to develop innovative solutions. • Performing both tactical competitive analysis (product-to-product) and strategic competitive analysis (company-to-company).
Infant Mortality	• Failure of a new product at the early stages of use by the customer. • It can be a metric of the rate of early failure of a new product. • Not to be confused with DOA (Dead on Arrival), which is a different metric.
Innovation	• The introduction of new capabilities that create value for customers. In the business world, innovation is so only if the customer says it is — only if the customer sees value in your offering. • The ability to deliver new value to customers across the entire set of customer experiences — from sales, to delivery, to support, operations and beyond. Every "touch point" with customers provides ample opportunity for innovation and thus value creation. • It's one thing to think up an innovation; it's another to develop it and deliver it to the customer. • Innovation is clearly not limited to technology!
Intangibles	• Attributes of an offering that can't be perceived by the senses. Goodwill, brand, image and position are examples of intangibles. • The role of intangibles in the customer decision process is often underestimated leading to products that are perceived by the customers as incomplete.
Intellectual Property (IP)	• Information, formulas, designs and other knowledge that is unique to an individual or company and can be used to grow a business or improve the competitiveness of an organization. • IP takes many forms, but the main criterion is that it can provide a competitive advantage or can be sold to someone who values it.

Glossary

Term	Definition & Notes
Issue	• A known obstacle, problem or anything that has 100% probability of occurring. And when it does occur, it will cause serious or fatal disruption to a program or project.
Kaizen	• A Japanese term that represents the philosophy of continuous improvement. • More than a set of tools or processes, Kaizen is a state or mind, a commitment to continually improve and striving for perfection.
Kano, Kano Analysis	• A model developed by Dr. Noriaki Kano to characterize quality as perceived by the customer. • Generally shown as a two-dimension diagram, the model maps the degree to which the new product performs versus the degree of satisfaction of the customer. • The model assumes that all input shown was provided by the customers, and not internal personnel. • The model can be very tedious and involved.
Lagging Metrics	• See "Outcome Metrics."
Lead Customers	• Carefully selected representatives of customer groupings (segments) that have needs and requirements typical of the rest of the segment. • In some companies, Lead Customers represent customers who are facing problems ahead of the rest of the segment, providing early visibility of future customer needs.
Leading Metrics	• See "Process Metrics."
Lessons Learned	• The benefit of performing retrospects (postmortems) and implementing the improvements as a result of the process. • Lessons learned do not benefit the organization until they are applied to improve the future performance of the organization.
Life Cycle Costs	• The total costs that the customer will incur during the useful life of the new product. These costs include purchase, maintenance, repair, installations, training operations and disposal. • Generally used during the selling cycle. Customer wants to know the total cost of a new product, not just the selling price.
Market Segment	• A group of customers who display similar characteristics relevant to a business. • Market segmentation should answer three questions: – Who are the customers? – Where are they located? – What are they trying to achieve?
Market Segmentation	• The process of grouping customers with similar characteristics. • Generally the segments represent the view of the customers; that is, how they classify themselves. • A market-driven segmentation should include your products or service. It should focus on how customers group themselves. Subsequent versions of the segmentation may include your products.

Glossary

Term	Definition & Notes
Market Sensing	• Research, observations and experience develop a deep understanding of the forces acting on a market to enable the development of a vision of the future. • Sensing the market is generally the responsibility of the senior staff and involves deep participation in the affairs of a market place. • Gathering customer needs and customer requirements is not part of Market Sensing.
Market Share	• A company's sales in a product area expressed as a percent of the total sales into that market by all competitors • Market share depends on what the company defines as its market and its competitors. • If it only considers competitors as companies that offer the same product type, it will have a higher market share than if it defines its competition as including substitution products.
Market-Driven	• A commitment to serve the needs of a market. • A chosen strategy and cultural shift that enable the achievement of a set of business objectives by focusing on serving specific market segments, gaining profound insight into specific market segments, delivering superior value and earning customer loyalty.
Marketing Requirements Document (MRD)	• The process that the marketing function uses to communicate to engineering (R&D) the problems that a new product must address. This takes the form of a document called the MRD, which articulates the prioritized customer requirements. Engineering uses the MRD to generate product concepts, features and specifications • The MRD goes by various names depending on the company or industry. There is always a document that serves to communicate the prioritized customer requirements to the R&D organization.
Mean Time Between Failures (MTBF)	• The number of failures in a given period of time divided by the total uptime in the same period. • A measure of reliability that represents the average time between failures.
Mean Time to Repair (MTTR)	• The total estimated downtime of the new product for a given period divided by the number of failures or repairs performed in the same period. • Used to plan maintenance programs, calculate Total Cost of Ownership and compare reliability.
Metrics	• A standard of measurement. For example, metrics will directly measure our progress toward achieving the organization mission and group's objectives. • There are three kinds of metrics: 1. Results metrics. 2. Process metrics. 3. Customer satisfaction metrics.

Glossary

Term	Definition & Notes
Mission	• The specific task or assignment adopted by an entire group or team. The Mission describes what it does, for whom it does it and the reason it does it. • The mission statement should be specific, achievable, easy to articulate and motivating. For example: – The mission of the police is to protect and to serve. – The mission of the product team is to get a specified product out on time and meet customer requirements.
Need	• Describes what the customer says he is lacking, what he would like to do or obtain or what he wants to get. • Customer needs, as expressed by customers, are often limited to product "features" (attributes) or problems that may or may not be related to the customer end goals or to how the decision will actually be made between competing offerings. It is the reason we introduced the term "Compelling Reason to Act."
Net Present Value	• Discounting future cash outflows and inflows using today's "cost of money." • A way to normalize several investments so you can compare apples to apples. • A way to determine the viability of an investment objectively. • Used to evaluate several alternative new products and compare their financial viability.
New Product Development (NPD)	• The processes, models and tools to develop a new product. • NPD ensures speed, repeatability and concurrency. • The processes must be documented, measured and improved in order to yield long-term benefits.
New Product Introduction (NPI)	• The process of rolling out or launching a new product to the market. • The go-to-market strategy and its corresponding tactics.
Objective	• A specific, definable and measurable end-result, which is to be pursued by a team or an individual of a company within a specific period of time. • Objective and Goal are often used interchangeably. See this definition.
Offering	• A set of products, services and intangibles offered to satisfy particular customer problems. • Express what the product is from the customer's perspective. • See also "Complete Product" and "Whole Product."
Outbound Marketing	• The function that leads all activities that take the products, programs and messages to the market. • Includes managing the distribution channels, enabling the direct sales channel in all its forms, marketing communications and customer retention programs, pricing and pricing schedules, new product roll-outs, etc.
Outcome	• The measurable outputs of an action project or program. The measurement of the actual results. • It is important not to confuse "completion" of a task with "results"; they are not necessarily the same.

Glossary

Term	Definition & Notes
Outcome Metrics	• Also called "Lagging Metrics," they track the final outcome or result of a project or process.
Pipeline	• See "Roadmap."
Plan	• A detailed formulation of a program of action; for example, a Marketing Plan. • A written set of tasks, schedules and resources with defined end dates and an assessment of any barriers which might put the schedule at risk. • There are many outlines for a plan, but generally a plan answers seven critical questions: 1. What needs to be done? 2. Who will do it? 3. By when will it be done? 4. How much will it cost? 5. What are the threats and risks? 6. Why are we doing it? 7. How will we mange change?
Policy	• A definite course or method of action to guide and determine present and future decisions. • An envelope for action with clearly defined boundaries. • The organization may choose to make policy inviolate to provide a more robust framework for action.
Portfolio Management	• A process to prioritize development projects based on a set of documented criteria. • Can be used any place where a business decision must be made. • The intent is to maximize the value of the portfolio to the corporation and to ensure that corporate strategies are in full alignment with the flow of new products. • Other names include Pipeline Management, Product Pipeline, The Product Funnels and The Product Portfolio.
Postmortem	• A process to understand the effectiveness of a project and to use the information to improve future performance. • Postmortems can be done at any time during the life of a project; they are most commonly done at the conclusion. • Postmortems are only useful if the information is used to improve the future performance of the organizations. • Postmortems are also known as Debriefs, Retrospects and Lessons Learned.
Procedure	• A written series of steps followed in a regular manner to accomplish a specific task in reproducible manner. • Similar to a process, but generally assigned to a single individual. Some compare a task with an action item.
Process	• A documented, sequential list of value-added steps that can replicate a product or service to a customer. • Unlike a project, processes are designed to be repeatable. A project is a single event with a clear start and end.

Glossary

Term	Definition & Notes
Process Metrics	• Also called "Leading Metrics," they measure activities that lead to a result. They are important to track progress towards a goal. These metrics are helpful to improve processes.
Process Out of Control	• When the number of defects exceeds boundary conditions. • Thus, an engineering organization, as a process, is out of control when there are no metrics to detect the number of defects nor a system for remedial and corrective action.
Product	• Anything that can be offered to a market that represents value; i.e., demand. • Good or service that meets the requirements of a market segment. • Colloquially, a manufactured good or the tangible part of value. But this is a limiting connotation, as customers in a business-to-business environment buy solutions to problems, and such solutions are made up of tangibles, services, experiences and intangibles. Sometimes the term "whole product" is used when referring to the entire package of tangibles, services, experiences and intangibles offered to a market.
Product Benefit	• The positive outcomes or results that an offering promises a customer. • Anticipated benefits apply before the purchase. Actual benefits apply after the purchase and manifest themselves as the customer's experience with your offering.
Product Development Team	• All the individuals whose contribution will result in the delivery of a new product to the marketplace. • In some organizations the Product Development Team is defined more narrowly; for example, only the marketing and engineering professionals involved.
Product Funnel	• See "Pipeline."
Product Life Cycle	• The four stages that all new products go through from introduction to end-of-life: introduction, growth, maturity and decline. • Duration of time from the earliest proposal to removal from the order entry system (phase out). A series of phases products follow from earliest conception to end-of-life • Product Life Cycle has many applications in product companies; such as tracking the product portfolio, ensuring profitability (gross margin time number of units), resource allocation, etc.

Glossary

Term	Definition & Notes
Product Management	• The process and strategy to mange the life cycle of a product such that it meets profitability targets. • The role of Product Management is to define, develop, deploy and maintain products and services that: – Provide more value than the competition. – Help build a sustainable competitive advantage. – Deliver financial benefit to the business. • Six of the key disciplines of product management are: 1. Market Research 2. Product Definition and Design 3. Project Management 4. Evangelize the Product 5. Product Marketing 6. Product Life Cycle Management
Product Manager	• The person assigned the responsibility and accountability for leading all the various activities that concern a particular product along its life cycle. • Sometimes "Brand Manager", particularly in consumer products. • Generally the Product Manager has P&L (Profit and Loss) responsibility for the product line. Some companies do not give P&L responsibility, but assign a "pro-forma" P&L to signify the importance of the job. • The Product Manager is responsible and accountable for achieving the business objectives of the product line.
Product Portfolio	• See "Pipeline."
Product Requirements Document (PRD)	• See "Marketing Requirements Document."
Program Evaluation Review Technique (PERT)	• A form of network analysis used to estimate the duration of a project.
Program Management	• The business function that assures the clean execution of corporate strategies, where a program is a group of related projects managed in a coordinated way to obtain the benefits and control not available from managing them individually. • The investment in program management is made to: – Reduce the variability of return from an investment. – Ensure alignment between strategy and execution in order to meet the business objectives of the enterprise. – Deal with change in a dynamic environment to ensure business results. • Program Management is not the same as "Project Management."
Program Manager	• A person with the skills, knowledge and abilities to lead a cross-functional team in the development of a product with a successful outcome such as profitable time-to-market. • This definition applies to the context of product development — your organization may use this title differently within a different context.

Glossary

Term	Definition & Notes
Programmatic Customer Visits	• A structured set of customer interviews where the participants share a common set of topics to be discussed. Further, the participants agree to a regular schedule of trips to the field to conduct the interviews. • A common way to capture, prioritize and archive the information from the customers.
Qualify Function Deployment (QFD)	• Quality Function Deployment is a structured methodology that uses a set of matrices for linking market requirements to the entire product development effort. • QFD will not define a product for you; it is simply a tool to manage information about the new product so that the integrity of the customer requirements is not jeopardized. • QFD is also called the House of Quality due to the appearance of the matrices. • QFD originated in Japan as a way to manage all the features and their interactions of a new product from start to finish.
Quality	• Perception of the degree to which a product or service meets customers' expectations. • Quality is whatever your customer says it is. • Quality is an attribute of the product that generates customer satisfaction. But it is also a perceived attribute, and therefore is somewhat subjective. • Quality Standards are approved, documented rules that define one or several dimensions of quality for a particular application or set of customers.
Rapid Prototyping	• Any of a variety of processes that avoid tooling time in producing prototypes. Generally these are nonfunctioning. • Rapid Prototyping exists to test the feasibility of a particular design with customers to reduce time-to-market.
Remedial Action	• Activities to provide an immediate resolution in order to solve a problem to the satisfaction of the customer. • Without regard to permanent solutions.
Request for Proposal (RFP)	• A process used by larger companies, government entities, military and universities to invite suppliers to bid on a specific acquisition of products, services, maintenance, etc. as a package. The submitted RFPs are then used by the buying entity to evaluate the merits and make an informed decision to purchase.
Request for Quotation (RFQ)	• See "Request for Proposal (RFP)."
Requirement (Customer)	• A brief and clear statement expressing what the customer will be able to experience as a result of owning and/or using the offering. • Product and customer requirements are two ways to express the same thing. • Requirements must be short, unambiguous, and able to be verified
Requirement (Product)	• A brief and clear statement expressing what an offering will have such that the needs of the customer will be met. • Product and customer requirements are two ways to express the same thing. • Requirements must be brief, unambiguous, and able to be verified.

Glossary

Term	Definition & Notes
Results	• The measurable achievement of an endeavor; for example, the results of a winning product is the achievement of the profitability target. • The outcome of an endeavor.
Rev (Review or Revise)	• To review or revise, used in software development to track the number of revisions that have been made to a specific application. • Revision can also be applied to documents to track version control.
Reverse Engineering	• The process of copying a product by analyzing every part of a competitor's product and then incorporating its features into one's own products. • Generally not a good thing to do.
Risk	• Something that can harm your program and should not happen. • The probability that something may go wrong with your project. • There are many types of risks that can interfere with your product development.
Risk Management	• The process to identify, prioritize, plan and manage risks in a project. • Risk Management starts at the earliest stages of a project and continues through completion.
Risk Management Plan	• The documented set of prioritized risks and corresponding mitigation plan. • The Risk Management Plan is constantly monitored.
Roadmap	• The manifestation of the product strategy for a product line or a company. • The outline of the future products versus the time it takes the organization to develop to achieve its business objectives. • Organizations may use terms like "Funnel", "Product Funnel", "Product Roadmap", "Pipeline" or "Product Portfolio." We suggest all of these terms are synonymous.
Simultaneous Engineering	• See "Concurrent Engineering"
Situation Analysis	• The current snapshot of the factors confronting a team, company or organization in a complex situation. • A short list of items, observations or data that are highly relevant to the topic at hand and are presently important to consider. • Examples include the Situation Analysis of a market and the Situation Analysis in a Product Proposal.
Sponsor, Team Sponsor. Team Mentor	• An Executive of the company who agrees to support a team in order to ensure its success. • The sponsor must have the time, willingness and skills to assist the team in a measurable way.
Strategic Alignment	• The endeavor to align all stakeholders on a single strategic direction such that their day-to-day activities fully support the endpoint. • Strategic Alignment complements (but is different from) "Strategy Formulation" or "Strategizing", which is the process of synthesizing and articulating a particular strategy.

Glossary

Term	Definition & Notes
Strategic Formulation (Strategizing)	• The act of formulating a strategy, different from the process of strategic planning. • Strategizing is the thought process or brainstorming needed to synthesize a strategy given a clear understanding of the market.
Strategic Initiative	• A program that, when completed, will deliver visible results in clear support of an overall strategy. It's defined to clearly move the organization towards the achievement of the corporate or group strategic goal. • Strategic Initiatives possess the following attributes: – Deemed critical to the success of the organization — non-negotiable and successful implementation is the only outcome. – In general, assigned to a member of the executive team as part of his MBO. – In general, implementation is carried out by one or more small teams that are given a clear charter, including deliverables, metrics, desired outcomes, time lines, roles and responsibilities. The executive is still accountable for measurable results. – Periodic reviews are included to ensure progress and accountability. – Fulfilled through programs and projects. – Span more than one fiscal year or are part of the yearly operational plan.
Strategy	• The approach that a team or company chooses to achieve a set of business objectives. • The approach guides the choices, behavior and direction of an organization.
Strategy (Corporate)	• A set of Strategic Initiatives made up of programs and projects that, when implemented, will fulfill the vision of the company. • This is the means by which you connect everyone's day-to-day activities to the vision of the company.
Strategy (Year Around Planning)	• A model of strategic planning that, unlike the yearly cycle of "business planning", assumes that strategy is a constant endeavor. • Usually characterized by a rigorous review and update of the strategy twice or three times per year.
Superior Value	• Better than your competition. Always defined by your customers, based on their comparison of you with your competitors. • Superior Value shows when a customer selects your products or solutions over that of your competition.
Switch-Over-Cost	• A term used to describe the mental process to shift one's cognitive resources from one topic to a completely different one. • Research done by Pattison, Kermit and Worker in "The Cost of Task Switching" shows that it may take up to 23 minutes to return to the same level of concentration after an interruption.
Team Charter	• See "Charter."
Team Sponsor, Team Mentor	• See "Sponsor."

Glossary

Term	Definition & Notes
Theory of Inventive Problem Solving (TRIZ)	• A Russian acronym that refers to a methodology for solving a difficult problem whereby multiple alternative solutions are defined. • The method is supposed to improve creativity by enabling the team to think of many solutions to a single problem. Generally a multi-discipline methodology to tap of the creativity of different individuals.
Timeline	• The schedule or the amount of time plus the activities needed to deliver a set of results. • A Gantt chart showing the chronological progression to the completion of a project.
Time-to-Market	• A measure of the speed by which a new product is developed. • Measured from the official start of the development until the product is ready to be purchased by a customer. • A critical measure to ensure competitiveness and success in the market place. • Various companies use different start and end points to calculate their time-to-market.
Time-To-Results (Time-To-Profit or Time-To-Cash)	• That point in the product life cycle where a target result that determines the success of the product has been met. • The success of a product may be measured by margin, revenue, market share or other strategic goals. • The time-to-results point is determined no later than the creation of the product proposal.
Total Cost of Ownership(TCO)	• The aggregate of all the costs during the life of a product, from acquisition to disposal.
Use Case	• Describes a specific activity or set of activities performed by a customer with a device or product. • Use cases are created with specific goals to understand the interaction between an operator and the equipment.
Value	• The customer's perception of what he wants to happen in a specific use and situation as the result of using a company's offering. • There is value in using the product or service — i.e., deriving a benefit as a result of use. There is also value in owning the product or service — i.e., deriving an emotional benefit just from owning the product or service.
Value Proposition	• A statement describing what resulting experience the customer will have, articulated in a way that communicates clearly and briefly the benefits, costs and competitive advantage of a particular offering. • The value proposition must be compelling and differentiated as perceived by the customer. • The resulting experience includes the outcomes or results that the customer will experience. • Whole products deliver value propositions.

Glossary

Term	Definition & Notes
Values	• A set of deeply held beliefs that enable employees to act without the need to seek management approval. • Values are never intentionally compromised. • A common set of social principles that guide the behavior and decisions of each person and the organization.
Vision	• The vivid description of the end destination and results of a group effort. • The distant lighthouse that provides direction to the completion of the team's journey. • A vision statement evokes concrete images. It is compelling and prompts forward progress. It is succinct and memorable.
Voice of the Customer (VOC)	• A process that uses structured, in-depth interviews for eliciting customer requirements. • VOC is also used as a generic way to talk about customer needs or customer requirements. • VOC by itself does not yield customer needs; other processes have to be used in conjunction to process the input from the customer, turn it into product features and eventually into a new product.
Voices	• Verbatim quotes from customers that provide insights as to their needs. • These voices are processed to extract requirements.
Whole Product	• The set of product, services, ancillary parts and intangibles needed to ensure that the customer will achieve his compelling reason to act (CRTA). • The actual product is a subset of the whole product. The whole product is what the customers buy. • Ignoring dimensions of the whole product explains why so many companies fail to penetrate new markets. • See also "Complete Product" and "Offering."

Life Cylces Models

New Product Development (NPD) Processes

Many processes that have been created to develop new products, all aimed at achieving these goals:

1. Reduce time to market.
2. Meet customer requirements.
3. Achieve profitability goals.
4. Deal with the variability (chaos) of the market place.

The journey to improve product development continues as the world changes. More and more, products are being developed in several sites around the world simultaneous. Outsourcing product development is a current trend, but one that has not gained wide acceptance. The use of vendors, suppliers, contractors and consultants is yet another trend. New trends for NPD will continue to emerge, which competitive pressures demand. And you need to deliver financial results in the midst of chaos.

Major Methods

Method	Description
1. Ad Hoc	No process or framework. The NPD is created as the development progresses. Not desirable as it does not provide the opportunity for continuous improvement. It is the most unproductive of all approaches. It works for a startup company, where there is generally a single product to develop and the team is relatively small, which helps communication and decision making.
2. Waterfall Model	A software development model first proposed by W. W. Royce in 1970. Royce presented it as a flawed, non-working model. It's a sequential development model in which each phase is completed before the next one starts, like a series of waterfalls, from requirements, to design to implementation, to testing and integration. This model is similar to the Phase Gate model for hardware development. Modern software development is much more dynamic than the Waterfall model allows; consequently, it has fallen into disfavor. Nor is it recommended for chaotic environments where customer requirements keep changing. It works best when the requirements and features of the new product have been defined early in the process, and little or no change is expected during the development phase. Then, the entire development process can be defined with a high degree of certainty.
3. Spiral Model	Developed by Barry Boehm to address the weaknesses of the Waterfall model, the Spiral model focuses on addressing risks incrementally by repeating the Waterfall model in a series of cycles or rounds. It accommodates multiple rebuilds and provides several opportunities to meet customer requirements. Best suited for projects where the scope is vague. It relies on the Do-Check principle: complete one section, check with the customer and then define the next phase. Somewhat similar to the Rolling Wave planning in project management, where you plan the next phase in great detail, but the rest is left vague by design. Used for projects where you are developing the new technology and at the same time the next product from such technology. This methodology requires considerable discipline and much maturity on the part of the design team. There is a risk of the endless design where a myriad of iterations take place, but no new product emerges.

Life Cylces Models

Method	Description
4. Iterative	Very similar to the Spiral model, and the start of the Agile approach for software development. Extreme Programming is another member of this category. Iterative development is a "rework scheduling strategy," in which time is set aside to revise and improve parts of the system.
5. Incremental	This model involves developing various part of the product at different rates and then integrating them as they are completed. Iterative and Incremental were merged in the 1990s through the Rational Unified Process (RUP).
6. Agile (Flexible)	The Agile approach promotes the interactions between the development team and the customer through the life-cycle of the development project. This model requires frequent access to customers and is best suited for IT environments. Characterized by developing software in short amounts of time, from one to four weeks, where each interaction is defined as a software project. Although more applicable to software development, some hardware companies are trying it. It's best suited for custom products, not for products aimed at main customers. Highly recommended when the customer requirements are neither firmly defined nor prioritized. Generally, the definition phase of the project takes longer than projects with more clarity.
7. Concurrent Engineering (Simultaneous Engineering)	The methodology assumes that many activities during the development process can be started and managed concurrently. There is close coordination by the core team to ensure that no waste occurs, all functions are working in parallel, all activities are value-added. Well suited for rapid product development, it allows a core team to serve as the steering mechanism for development. The core team is made up of representatives of the major functions involved in the development such as engineering, marketing, manufacturing, service, finance, etc.
8. Phase Gate (Stage Gate)	Attributed to Robert G. Cooper, Phase Gate is a very common process in the hardware world, but not sufficient for fast-changing markets where speed is essential. Phase Gate involves a series of review gates along the product development cycle. These gates are generally associated with major milestones of the project. The development program is divided into a number of gates, and a certain criteria must be met before moving to the next gate. Versions like Hard Gate and Soft Gate have been created to help speed up the process. Phase Gate lends itself to a project management environment. Be careful that it does not become too bureaucratic and too focused on the gates themselves.

Life Cylces Models

Don't Be Fooled by Fads

What about software packages that promise to improve our produce development? Do not be fooled by new fads and models that promise much and deliver little. Some may even bring your efforts to a complete halt. There are a myriad of packages available and more are emerging every month.

1. It is critical that you and your team fully understand the current NPD and understand the obstacles and barriers that keep you from gaining speed prior to selecting any software package.
2. You and your team should also understand the characteristics of the new method you intend to deploy. Use the Web, books and other resources to equip yourself with the knowledge about the proposed methodology so you can make informed decisions.
3. We strongly suggest that you adopt new methods without any software. Take a year or two to become familiar with the method. After that, you can start evaluating software offerings.
4. All software providers will offer a very rosy picture of their software. No doubt, they believe in it, but also they want you to purchase their package. Be sure to evaluate all offerings objectively and thoroughly. Talk to current and former customers of the vendors directly. Visit web sites that evaluate and critique software applications.
5. Beware of customization. This may add considerable time and cost to your package. Have a clear understanding as to what is included and what is not. Avoid surprises by having this conversation prior to signing any contracts.
6. Be certain that the package includes user training, installation training and technical support at the very least. Any new software package will have a negative impact in the productivity of your team before any benefits are realize. You and your managers must be fully aware of this occurrence and ensure that focus during deployment is the training, coaching and support of the users.
7. Finally, remember that any new software application must fit the business objectives of your company. Be compatible with the current corporate culture or be prepared push back.

Bottom Line

There are yet more product development models than we've listed, and new ones are emerging as well. The key message is that you can create or adapt a model to suit your business objectives and the culture of your organization.

Remember that every NPD process has four objectives: reduce time to market, meet customer requirements, achieve your profitability goals and deal with the chaos of the market place.

It is OK to experiment, but you and your team must clearly understand that it is an experiment, and you must have a clear reason for the experiment.

The best way to improve your product development process is to have formal debriefs (postmortems) and a robust process to deploy the improvements.

Templates

Instructions

These templates are for you to use and share. Throughout this guidebook, you have seen real world tools and examples. Any page with a template icon indicates that there is a correlating fillable form in this section.

For more resources and future updates, visit us at: www.rapidinnovation.com.

Templates with Examples

More Tips for Users of the Interactive PDF Version

You can view the complete list of templates with the "Show/Hide Templates" bookmark on the left Navigation Panel. This reveals the Attachments Tab at the bottom of the screen. Double click any template you want to work on. You can also open a template by double clicking the white pushpin icon on the specially marked example pages.

Once opened, these templates are enabled for use in Adobe Reader or Adobe Acrobat — templates that you can fill with your own information and share with your peers and colleagues.

When you open original templates from your guidebook, they will be fresh. But you can save versions for future access — whether you have completed the template, partially filled in your information or left in blank. Saved versions remain enabled for revisions and collaboration.

When you close the template, you will return to this guidebook.

These enabled templates also provide you with Comment & Mark-up Tools. Just as you might with a hard copy, you and your team can highlight text, add notes, and make custom annotations.

Product Value Profile

Rate each offering on a scale of 1-10 for performance in delivering the stated benefit or cost: 1 meaning "unacceptable," 5 meaning "satisfactory" and 10 meaning "outstanding."

No.	CVDs in Order of Importance	Category	1				5					10	Differentiation
1													**BEST** You need to be the best in satisfying the most important Customer Value Drivers (CVDs).
2													
3													
4													**COMPARABLE** You can be comparable to competitors in the second tier of CVDs.
5													
6													
7													
8													**INFERIOR** You can be inferior in satisfying CVDs that are the least important, provided your performance is still acceptable.
9													
10													

Legend: Company & Line Style

1. ______________________
2. ______________________
3. ______________________

Well-Differentiated Product: ________

A Product Value Profile graphically represents the ability of competing products to meet the most important Customer Value Drivers for a target segment. The "well-differentiated product" is one that satisfies the three most important Customer Value Drivers better than any competitor, while being comparable to them on the next tier and even inferior on the last tier.

Product Value Index

(A)	(B) Rank	(C) Customer Value Drivers in Order of Importance	(D) Category	(E) Your Product	(F)	Your Competitor's Product		
No.	Importance			Rating (1-10)	Score	Rating (1-10)	Score	Differentiation
1								**BEST** You need to be the best in satisfying the most important Customer Value Drivers (CVDs).
2								
3								
4								
5								**COMPARABLE** You can be comparable to competitors in the second tier of CVDs.
6								
7								
8								**INFERIOR** You can be inferior in satisfying CVDs that are the least important, provided your performance is still acceptable.
9								
10								
All		**Product Value Index**			(G)			

The PVI (Product Value Index) is a numerical expression of value. By quantifying the value that a product represents to a target segment, you can determine how the offering of one company ranks compared to that of another. The PVI is calculated by first multiplying the Importance percentage (B) by the Rating (E) — this generates the Score (F) for each CVD — and then totalling all the Scores to create the PVI (G).

In this example, Your Product has a PVI of 7.77. This score is relevant only in comparison with another PVI. The example shows that Your Competitor's Product has a PVI of 8.02. Though not a big difference, only one product can win in each situation. In addition, Your Competitor's Product has better ratings for the top value drivers, increasing its chance to win in side-by-side comparisons.

Charter Tool

1. Scope of the overall project — objectives for the new product or application:

- ❖
- ❖
- ❖
- ❖
- ❖
- ❖
- ❖

2. Purpose of this set of visits:

- ❖
- ❖
- ❖
- ❖
- ❖
- ❖
- ❖

3. The objective for the visits:

- ❖
- ❖
- ❖
- ❖
- ❖
- ❖
- ❖

4. Scope for the Voice of the Customer in this project:

- ❖
- ❖
- ❖
- ❖
- ❖
- ❖
- ❖

5. Key dates for the set of visits:

- ❖
- ❖
- ❖
- ❖
- ❖
- ❖
- ❖

6. Who leads this set of customer visits?

- ❖
- ❖
- ❖
- ❖
- ❖
- ❖

7. Who will pay for this set of customer visits?

- ❖
- ❖
- ❖
- ❖
- ❖
- ❖

8. Who will approve or disapprove the Marketing Requirements Document (MRD)?

- ❖
- ❖
- ❖
- ❖
- ❖
- ❖

9. Team members for this set of visits:

- ❖
- ❖
- ❖
- ❖
- ❖
- ❖

10. Sponsorship:

- ❖
- ❖
- ❖
- ❖
- ❖
- ❖

11. [Add your own item]:

- ❖
- ❖
- ❖
- ❖
- ❖
- ❖

The Four Critical Questions

Question	Examples			
Who The title or function of the person to interview (profile).				
How Many The number of interviews for each of the categories.				
Venue The forum you have selected to conduct the interview.				
Cross-Section The segments or groups from the chosen market.				

Customer Archetypes

Subsegments A and B are there to provide one additional level of segmentation. For example, you might make A the U.S. and B Europe or you can make A Managers and B Technicians. You need not use both A and B in your sample selection; it's your choice.

Types of Customers	Subsegment A	Subsegment B
1. Customers you have lost. *That is, customers who no longer buy from you for a negative reason.*		
2. Customers you never had. *That is, customers who ought to have bought from you, but never did.*		
3. Customers who have stopped buying from you. *That is, you have not lost them, but they have not bought from you in some time. Different than Item 1 above.*		
4. Future customers. *That is, customers in new markets, new companies, new applications, etc.*		
5. Customers who are buying much less from you. *Those who, in the last six, 12 or 18 months have shown a drastic reduction in purchases.*		
6. Customers who are buying much more from you.		

Types of Customers	Subsegment A	Subsegment B
7. Your competitor's customers. *While hard to identify, the effort is worthwhile. You will get a brand-new perspective on you, your products, and your company.*		
8. Your most profitable customers.		
9. Your least profitable customers.		
10. Customers who cost you money. *Customers who demand large discounts and expect added features and services. In some cases, you sell to them at a loss, even though that is not your intention.*		
11. Upset customers. *Customers who are upset at you but still buy from you.*		
12. Power users. *Those customers who are always ahead of the curve. They enjoy pushing the limits of technology. Sometimes called the early adopters; not the same as lead customers. They like to have the latest things, even though they may not need them.*		
13. Lead customers. *Customers who are developing new products that may revolutionize a market. Those that are creating breakthrough products. Not the same as power users.*		

Types of Customers	Subsegment A	Subsegment B
14. Channel partners. *Your distributors, VARs, REPs and other alternate channels of distribution. Naturally, also their customers.*		
15. Lost leads (almost customers). *A customer who expressed an interest in buying your product, but either did not buy or purchased a competitor's product.*		
16. Mix-and-match customers. *Customers who buy from you and your competitors. Typically they like your products, but they still buy from your competitors as well.*		
17. Distant customers. *Customers who are the farthest away from your office. Generally, they have a very different set of perspectives.*		
18. Define your own category.		
19. Define your own category.		

Decision-Making Tree

Represent the 80 percent case, not perfection, as there will always be exceptions. This approach focuses on how decisions are made in an organization. Each stakeholder has a set of needs. Use this approach to identify the needs of the different stakeholders; for example, if more than one person is involved in deciding the purchase of your product. This is particularly important if your company makes capital equipment that require many approvals.

Market Segment or Customer Group: ______________________________

Internal Influences

Anyone or anything within the customer's organization that can influence the purchase decision.

Final Approval

Person who has control over funding.

External Influences

Anyone or anything outside the customer's organization that can influence the purchase decision.

Key Decision-Maker

Usually the person responsible for the budget, head of the department or other type of senior manager.

End Users

The problem-owners.

Examples of internal influences:

- People who install.
- People who control quality.
- People who maintain and repair.
- People who repair or service.
- People who support and help.
- People who do the planning.
- People who provide the overall funding.

Examples of external influences:

- Other customers.
- Competitors.
- People who regulate and rule.
- Local industry trends.
- National industry trends.
- Global industry trends.

Classic Tree Structure

Layer #1

MAJOR CUSTOMER SEGMENT OR GROUP

For example, internally, a major department, or externally, a geography, business type, or industry.

Best when the segment is defined by the customers themselves; i.e., it is a self-referencing segment; so that when you ask customers to which industry they belong, they articulate the segment.

Layer #2

SUBSEGMENT OR SUBGROUP

These are the segments within the Major Customer Segment in Layer #1 on the left.

Layer #3

TARGET CONTACT

Enter the specific information about the individual (customer) you will be contacting. The more specific the better. For example, names, titles, or positions.

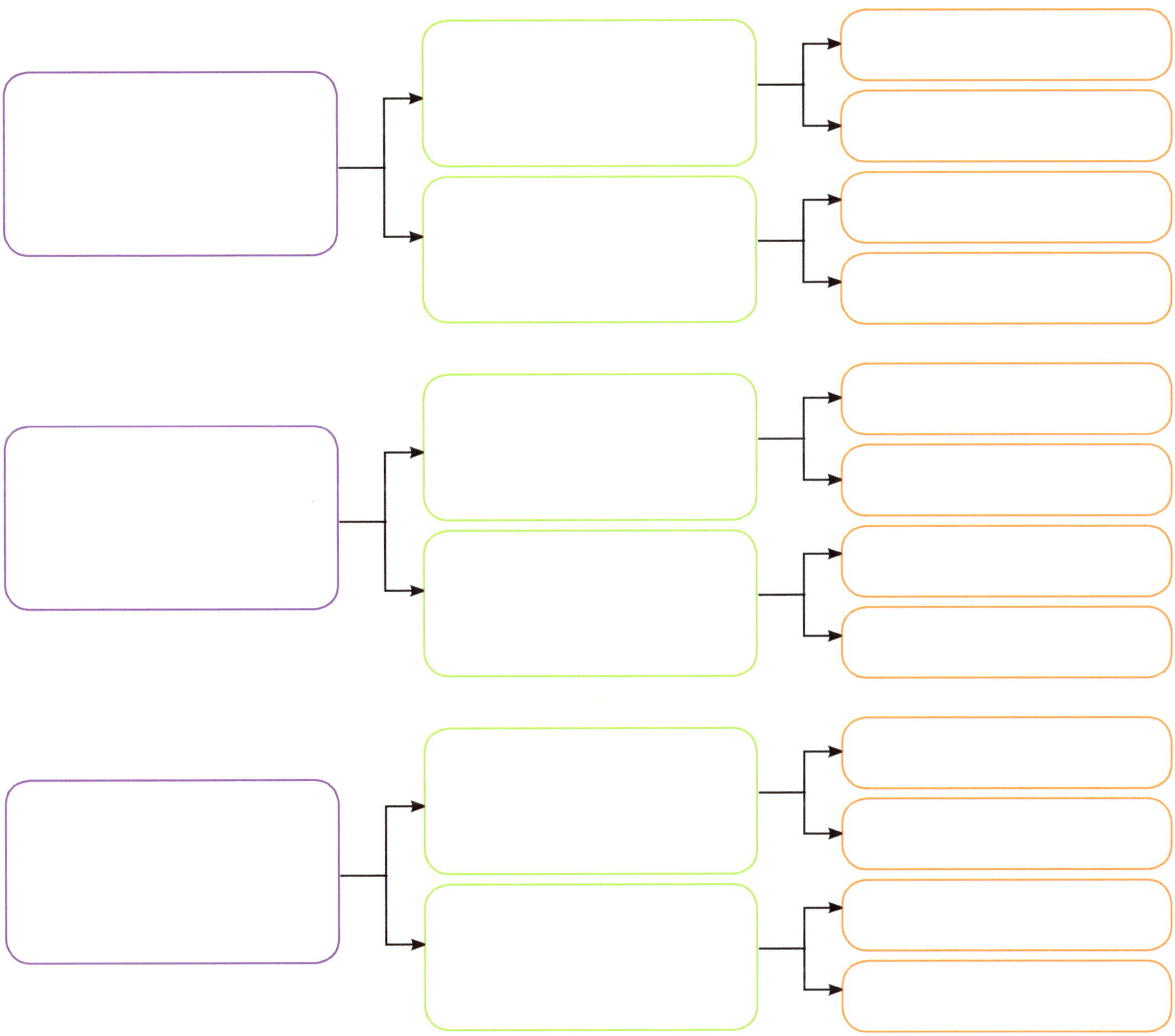

Market Segmentation

This sample selection approach uses the traditional market segmentation model. It is important to note that you still need to identify an individual to interview; that is, the title or name of an individual

Term	Market #1	Market #2	Market #3
1. Name of the market segment. *Could be by geography, application, industry, etc. Use the segmentation readily accepted by your organization.*			
2. Market share of units or dollars sold in each segment. *The percent share that you own of the total number of units sold into the selected segment.*			
3. Function or title of the person who is the primary user. *This is critical for sample selection. Without titles it is very difficult to trace or identify the proper person to visit.*			
4. What does the primary user do with the product?			
5. Function or title of the person who makes the purchase decision. *This is critical for sample selection. Without a title it is very difficult to trace or identify the proper person to visit.*			

Determining How Many Interviews to Do

Interviews	Type of Development Program		
	3 or more (maximum 12)	5 or more (maximum 12)	7 or more (maximum 24)
	Minor modification to an existing product — for example, an A version or a maintenance release	A new product from an existing platform	A new platform aimed at a new market dramatically different from your current products
Individuals from a single company with many sites			
Individuals to interview per each site			
Functional types — i.e., the title of the customers to be interviewed			
[Define your own category]			

Mapping Visits for Relevant Cross-Section

Group Dimension and Customer Archetype

Group	Customers					
Sample Selection Sub-Groups	Current satisfied customers	Current non-satisfied customers	Lost or competitor's customers	Customers gained from competitors	Not yet customers	Total
Total Interviews						

Cross-Check Visits for Relevant Cross-Section

Target Company and Function

Visit ID#	Target Company	Target Function (Who)	Venue (Where)	Number of Interviews	Cross-Section Criteria
Used for tracking purposes	Name or type of company	Name and title	Where the interview will occur	Interviews with this profile	How the selected criteria is met

Contact Information Checklist

Our Information

MY NAME: ______________

INTERNAL CONTACT #1: ______________

Phone #1: ______________

Phone #2: ______________

E-mail: ______________

INTERNAL CONTACT #2: ______________

Phone #1: ______________

Phone #2: ______________

E-mail: ______________

THE REST OF OUR TEAM: ______________

KEY DATES: ______________

Special Factors

POSSIBLE POSITIVE OR NEGATIVE INFLUENCES ON THE DIALOG WITH THIS CUSTOMER: ______________

BUYING HISTORY: ______________

❑Their Annual Report ❑Brochures
❑Product Data Sheets

COMPETITIVE SITUATION: ______________

CUSTOMER'S APPLICATION: ______________

Customer's Information

COMPANY: ______________

Phone #1: ______________

Phone #2: ______________

E-mail: ______________

Web Site: ______________

ADDRESS: ______________

City: ______________

State: ______________ ZIP: ______________

DIRECTIONS: ______________

OTHER NOTES: ______________

Key Contacts at the Customer's Site

CUSTOMER #1: ______________

Title: ______________

Department: ______________

Phone #1: ______________

Phone #2: ______________

E-mail: ______________

Other: ______________

CUSTOMER #2: ______________

Title: ______________

Department: ______________

Phone #1: ______________

Phone #2: ______________

E-mail: ______________

Other: ______________

TODAY'S DATE: ____________

Contact Information Checklist (cont'd)

Pre-Visit Checklist

PREPARATIONS REQUIRED: ____________

OUR PROMOTIONAL MATERIAL: ____________

- ❑ Audio-Visual Presentation
- ❑ Demo Equipment, software, etc.

OTHER DETAILS: ____________

Objectives of the Visit

PRIMARY GOALS:

#1: ____________

#2: ____________

#3: ____________

Discussion Guide

BASIS FOR STARTING THE CONVERSATION:

QUESTIONS OR THEMES TO BE DISCUSSED:

#1: ____________

#2: ____________

#3: ____________

Anticipated Commitments

OURS: ____________

THEIRS: ____________

Travel Planner

Travel Agency/Agent: ____________________ Web Site: _________________ Phone: ______________

Other Notes: ____

____

SUN	MON	TUE	WED	THU	FRI	SAT

Visit Program Planner

Visit#	Team Members	Customer to Visit	Start	End	Contact On-Site	Notes

Hierarchy of Topics and Questions

This form must be completed before you talk to your customers. First-Level Topics: Start by selecting and prioritizing the key topics of discussion. A topic is not a question — think of it as an area of conversation. Write the topic or subject to be discussed with your customers. Second-Level Topics: Develop key questions that will address the topic. Third-Level Topics: Conclude by developing follow-up questions on the same topic.

First-Level Topics (Themes)	Second-Level Topics (Themes)	Third-Level Topics (Themes)
Key Topic #1		
Key Topic #2		
Key Topic #3		

Planning Checklist

Deploying a tool is indeed a project. Consequently, ensure that you have a deployment plan, a schedule, budget, etc. Working with your Core Team, or with the peers who have agreed to help, you can develop a tactical plan that will ensure a successful deployment.

The list below is intended to help you formulate your own plan. You may need to add your own items to the list. Also note that the list below is in no particular sequence. Finally, the list is not rigid. Use your judgment and do what makes sense within your organization.

Planning Checklist

- ❑ Schedule for the deployment defined, documented and approved.
- ❑ Training materials for employees developed and documented.
- ❑ Training materials produced — hard and soft copies — available for use.
- ❑ Training materials placed in the intranet or web page.
- ❑ Trainers of the new tool identified and informed of their duties.
- ❑ Training of employees scheduled and employees notified.
- ❑ Training of managers and supervisors scheduled.
- ❑ Feedback mechanism in place.
- ❑ Metrics to measure the progress and eventual success of the deployment identified and ready to use.
- ❑ Coaching and encouragement for managers, employees and stakeholders defined and ready — including those who will do the coaching.
- ❑ Support available; for example, facilitators, coaches, tech support, assistants, etc.
- ❑ Obstacle removal in place.
- ❑ Templates, checklists and other tools ready and available.
- ❑ Information on our web page done.
- ❑ Processes that will be impacted by the deployment identified and corrected.
- ❑ Changes to our quality manual completed.
- ❑ Manuals, guides and other reference materials that will be needed by the users ready and available.
- ❑ Executives briefed and ready to fully support the deployment.
- ❑ Human resources (HR) informed.
- ❑ Training department informed.

Planning Document

This template may help you secure assistance from your executives. It is also a good point to start the process of deployment. It may even become a standard template for deployment of future tools and methodologies. Modify this template to meet your specific objectives within your organization.

1. Tool or methodology to be deployed.
    - ❖
    - ❖
2. Objective of the tool deployment.
    - ❖
    - ❖
    - ❖
3. Who is the expert on this tool?
    - ❖
    - ❖
    - ❖
4. What documentation is available?
    - ❖
    - ❖
    - ❖
5. Are there any external resources available?
    - ❖
    - ❖
    - ❖
6. What are the risks involved in the deployment?
    - ❖
    - ❖
    - ❖
7. How will we pilot the tool?
    - ❖
    - ❖
    - ❖
8. What will be our communication plan?
    - ❖
    - ❖
    - ❖
9. What will we measure to assure us that the deployment has been successful?
    - ❖
    - ❖
    - ❖

Charter for a Customer Panel

1. What is the purpose of the Customer Panel? Why are we holding this meeting? For what intent?

- ❖
- ❖
- ❖
- ❖
- ❖

2. What are the objectives? What is the measurable accomplishment from the meeting?

- ❖
- ❖
- ❖
- ❖
- ❖

3. What are the tangible results that we must have from this panel; i.e., the outcomes?

- ❖
- ❖
- ❖
- ❖
- ❖
- ❖

4. What are the methods that we will use to achieve the objectives? What is the overall approach?

- ❖
- ❖
- ❖
- ❖
- ❖
- ❖

5. What are the roles for the panel?

- ❖
- ❖
- ❖
- ❖
- ❖
- ❖

6. What will happen the day after the panel?

- ❖
- ❖
- ❖
- ❖
- ❖
- ❖

7. What will happen a week or two after the Panel?
❖
❖
❖
❖
❖
❖

8. How will we evaluate the success of the Panel?
❖
❖
❖
❖
❖
❖

9. Who from our organization must be part of this panel?
❖
❖
❖
❖
❖
❖

10. Logistics?
❖
❖
❖
❖
❖
❖

11. [Add your own item]
❖
❖
❖
❖
❖
❖

12. [Add your own item]
❖
❖
❖
❖
❖
❖

FREE BONUS

How to Get Your Interactive Version of this Guidebook

Register Your Book

Thank you for purchasing *Risk Management and FMEA for Product Development*. Many find the interactive PDF version of this guidebook to be a valuable addition because the cross-reference links are live and all the templates can be filled out electronically and used to collaborate with your team members around the world. Once you enter the information, every template becomes a separate PDF file for sharing and collaboration. Furthermore, the electronic version of this guidebook can also be stored in any portable device that accepts PDF files such as tablets, laptops and even smartphones.

To obtain your free interactive PDF copy of *Risk Management and FMEA for Product Development*, follow these simple steps:

1. Send an email to voc@rapidinnovation.com with "**Registering VOC book**" in the subject line. In the body of your email, include your name plus the following registration code: **VOC4712E**.
2. In a few days, you will receive an email notice that your book has been registered with either an attached copy of your free interactive PDF or a link to download your copy.

If you have any questions during registration, please email the publisher at: voc@rapidinnovation.com.

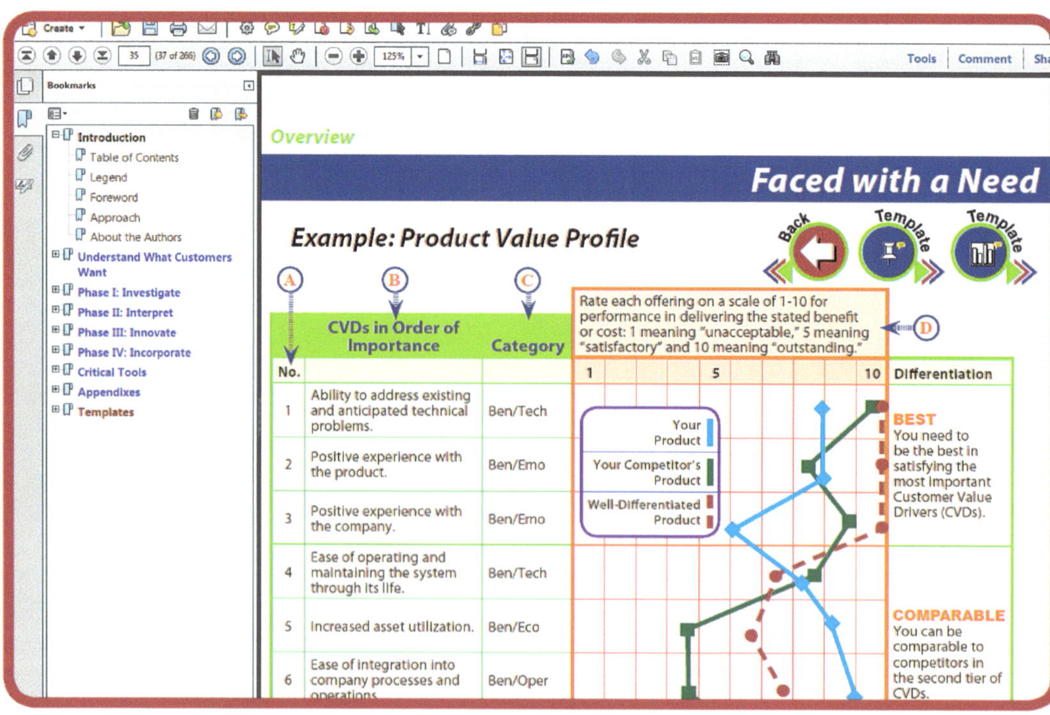

Other Titles You May Be Interested in Owning

THE SERIES

Flexible Project Management for Product Development

Traditional project management approaches are not sufficient for product development due to the complexity of getting a new product to market. Technology changes, personnel turnover, changing customer requirements, multi-site development, and unclear schedules are but a few of the variables that render conventional project management deficient.

Flexible Project Management for Product Development allows you and your team to manage your development project in the middle of the turmoil created by global competition. Beyond just reading very useful material, you will be able to use the templates and tools and also be able to share the information with your peers around the world. This is a step-by-step how-to guide. The authors realize that in the real world, you need more guidance and enough detail to enable you to apply this material to your specific development project. It starts by helping you to organize your development project to ensure that you build a robust environment to deal with chaos.

Risk Management and FMEA for Product Development

"If you are not managing risk, you are not managing your project." This is most applicable when it comes to managing risk in product development. Risk is the probability of something going wrong with your development project. Technology changes, customer requirements vary, schedules are always in flux, and your suppliers are in turmoil. In short, there are many things that can go wrong. What if you could predict what could go wrong? And more importantly, what if you could develop a plan to eliminate or mitigate your risks?

Risk Management is the ability to anticipate risks and develop plans to manage them; thus, creating more predictable schedules, improving reliability and customer satisfaction. In many cases, risk management prevents what otherwise might be a catastrophic failure of your development program.

This book was written by product developers with many years of experience in the "real world," where traditional text book approaches fail. This is a how-to guide that understands the pressures placed on you, the need to get to the point quickly and to show you step-by-step how to apply risk management on product development projects.

Debriefs and Postmortems for Product Development

Studies show that debriefs, or postmortems, are the best way to improve the productivity of teams — better than training, books, seminars and other tools. Your ability to methodically identify areas of improvement from development programs and immediately apply them to upcoming efforts will result in improved productivity, more enthusiastic teams and improved morale in your development organization.

This step-by-step guidebook will help you avoid the common mistakes many organizations make that block them from obtaining the full benefits from debriefs. You will learn how to prioritize improvement changes and how to deploy them through the organization in a way that it is embraced by all.

www.ingramcontent.com/pod-product-compliance
Lightning Source LLC
LaVergne TN
LVHW070949210926
848482LV00002B/147